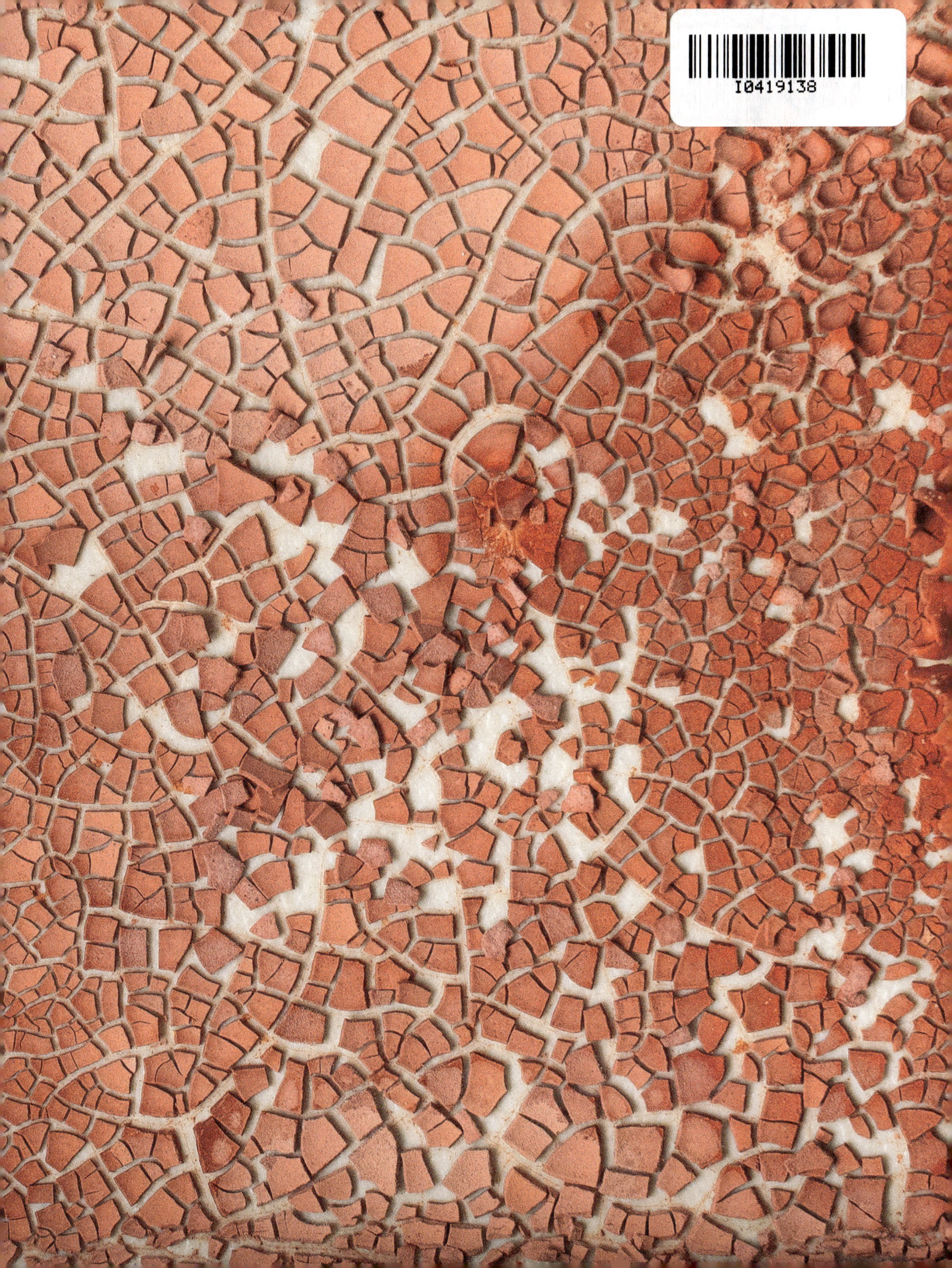

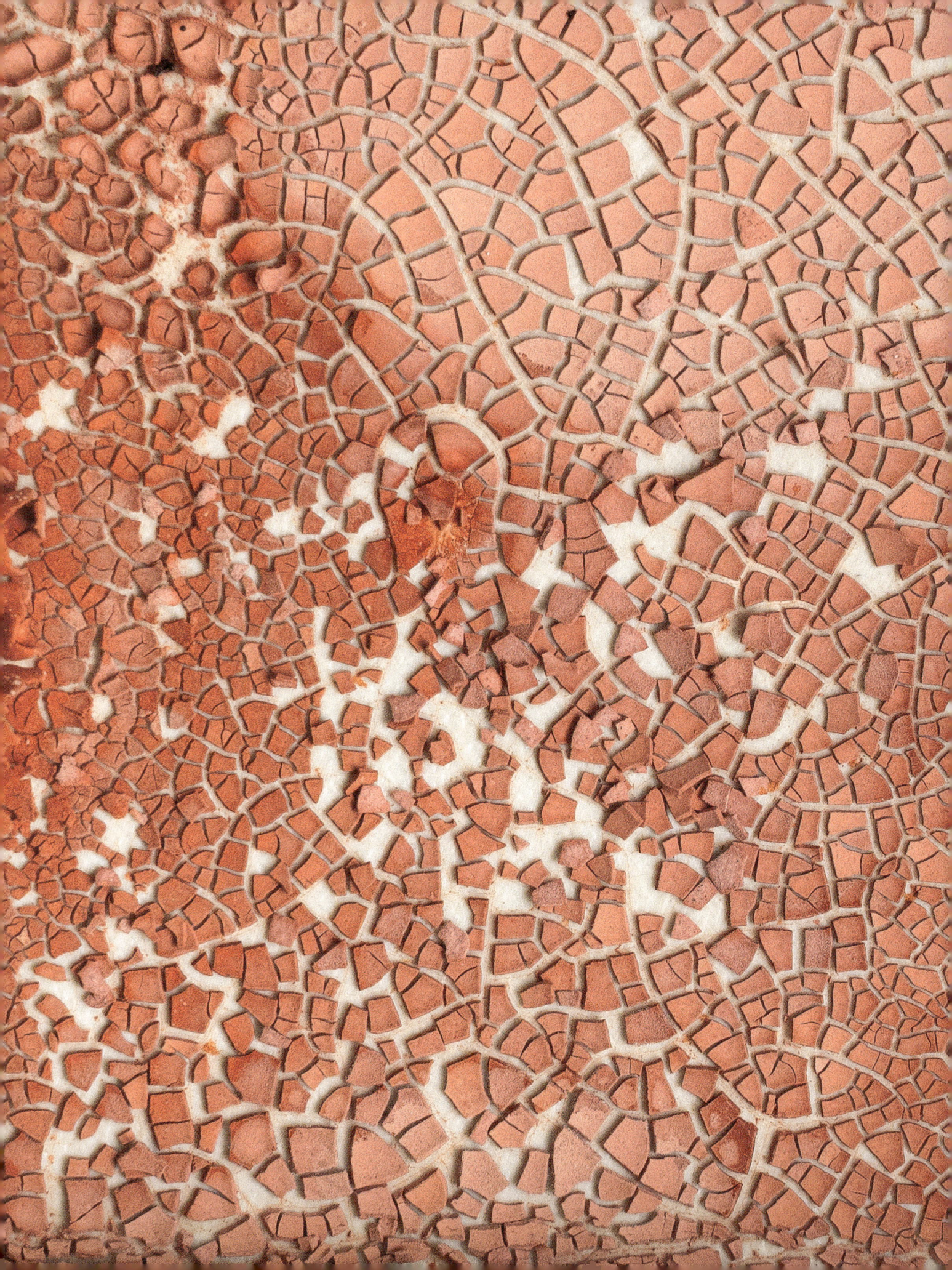

THE
NATURAL PIGMENT
HANDBOOK

A MAKER'S GUIDE TO THE ART, STORIES
AND RECIPES FOR CREATING PAINT

LUCY MAYES

DAVID & CHARLES
—PUBLISHING—

www.davidandcharles.com

CONTENTS

CHAPTER SIX:
THE FUTURE OF
PIGMENT PRACTICE

FOREWORD

I have written this book as a guide for those who want to delve deeper into their use of colour and pigment-making practice. I hope it will complement the amazing books that I have read on the subject and fill some of the small spaces that lie in between them. I've been composing it for a decade both in and out of my head, and so it's good to put into print some of the recipes, thoughts, and principles I have been teaching through my workshops.

Passionate about the craft of pigment making, I helped it receive endangered status on the *Red List of Endangered Crafts 2023* produced by Heritage Crafts (see Useful Organizations). It is my hope that my book might contribute towards the craft's progression and use. It is a snapshot of my experience as a pigment maker and artist and as such, it is deliciously incomplete, as I have much more to do and make. The material included here represents my education within an art university context in the Western Hemisphere, with a leaning towards pigment applications for oil painting techniques. The inclusion of pigment usage by other cultures is brief, although I hope that this will encourage further exploration and support of Indigenous knowledge.

I dedicate this book to all of the living organisms, people, plants, and animals that have been negatively affected by extractive processes in our human quest for the brightest and most beautiful colours.

Let's go back to earth.

Lucy

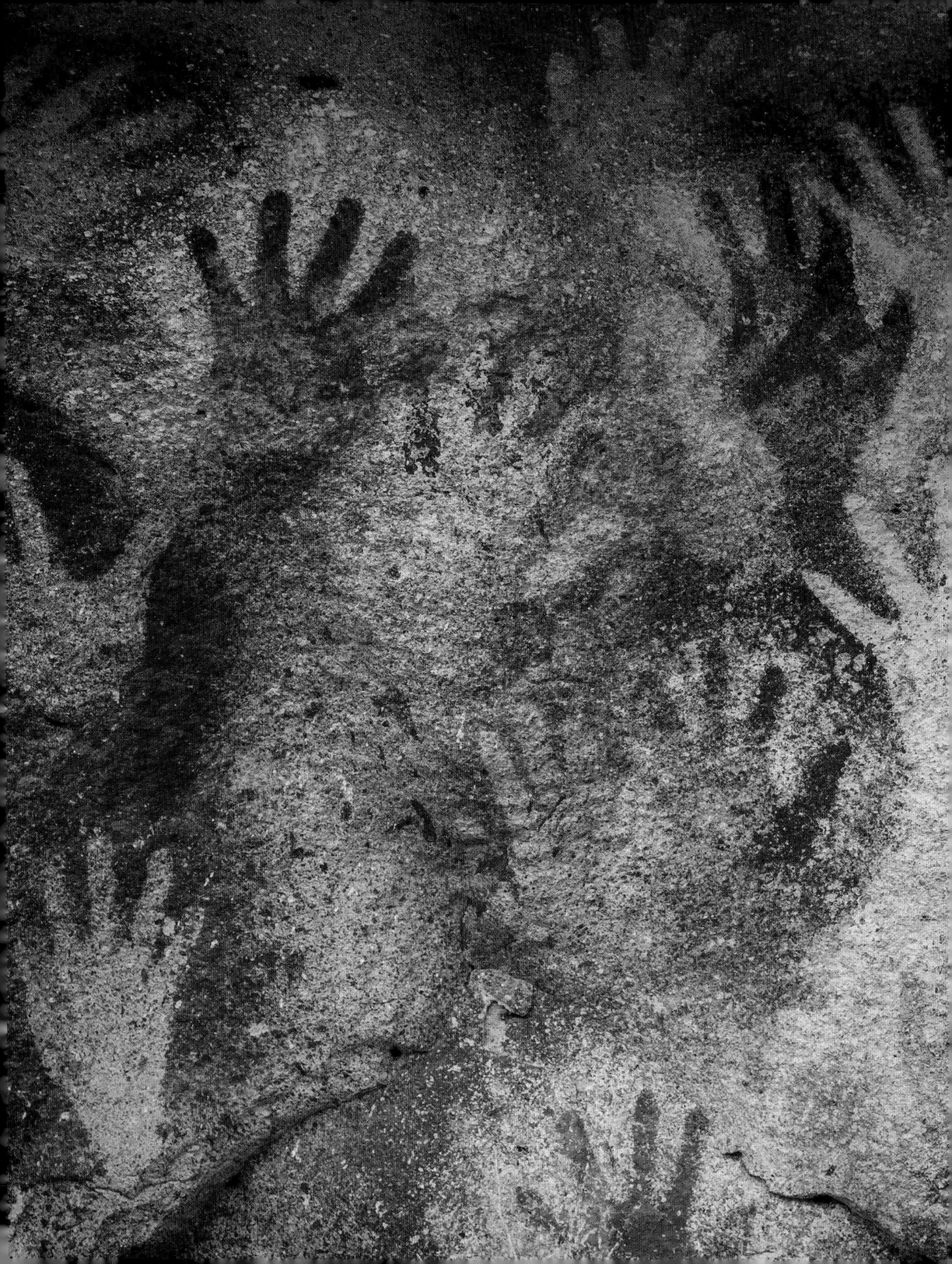

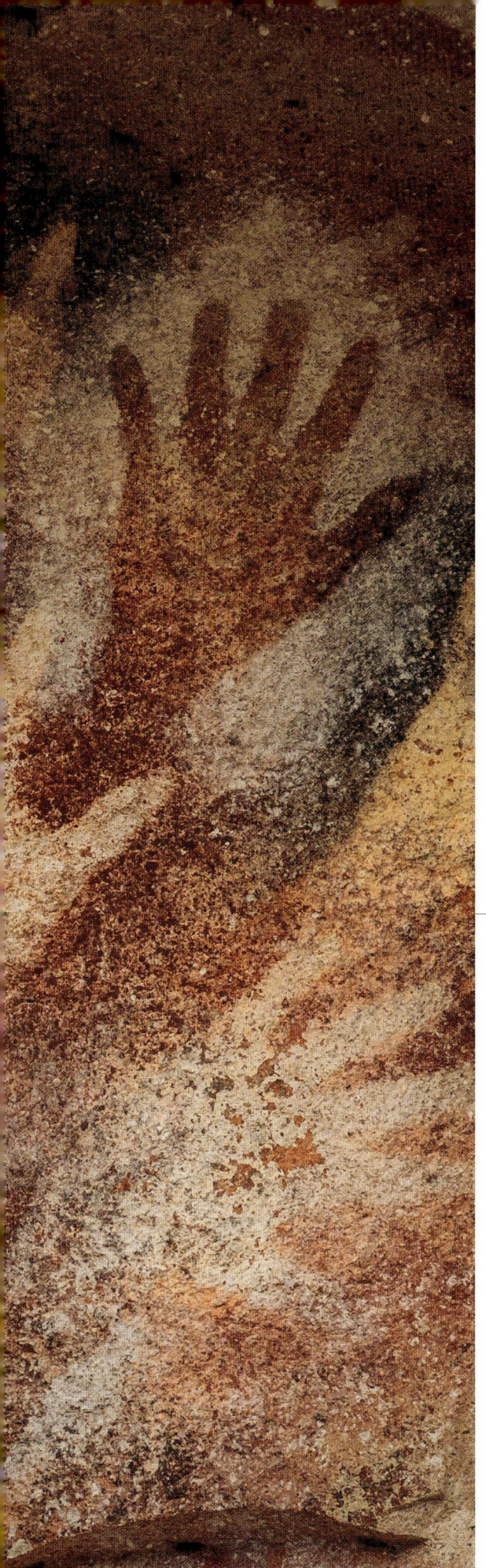

CHAPTER ONE

A HISTORY OF PIGMENT

In the following pages, we explore key insights and highlight important developments in the expansive history of pigment use and manufacture throughout the ages. Through detailed recipes for transforming earth through heat (calcination) and the creation of watercolour paints, you are invited to engage with the materials themselves. These rhythmic processes serve as a means of reflection, guiding you through the unfolding narrative of the text.

Left: Handprints at Cueva de las Manos (Cave of the Hands), Santa Cruz Province, Argentina, created between 7,300 BCE and 700 CE.

WESTERN PIGMENT USE THROUGHOUT TIME

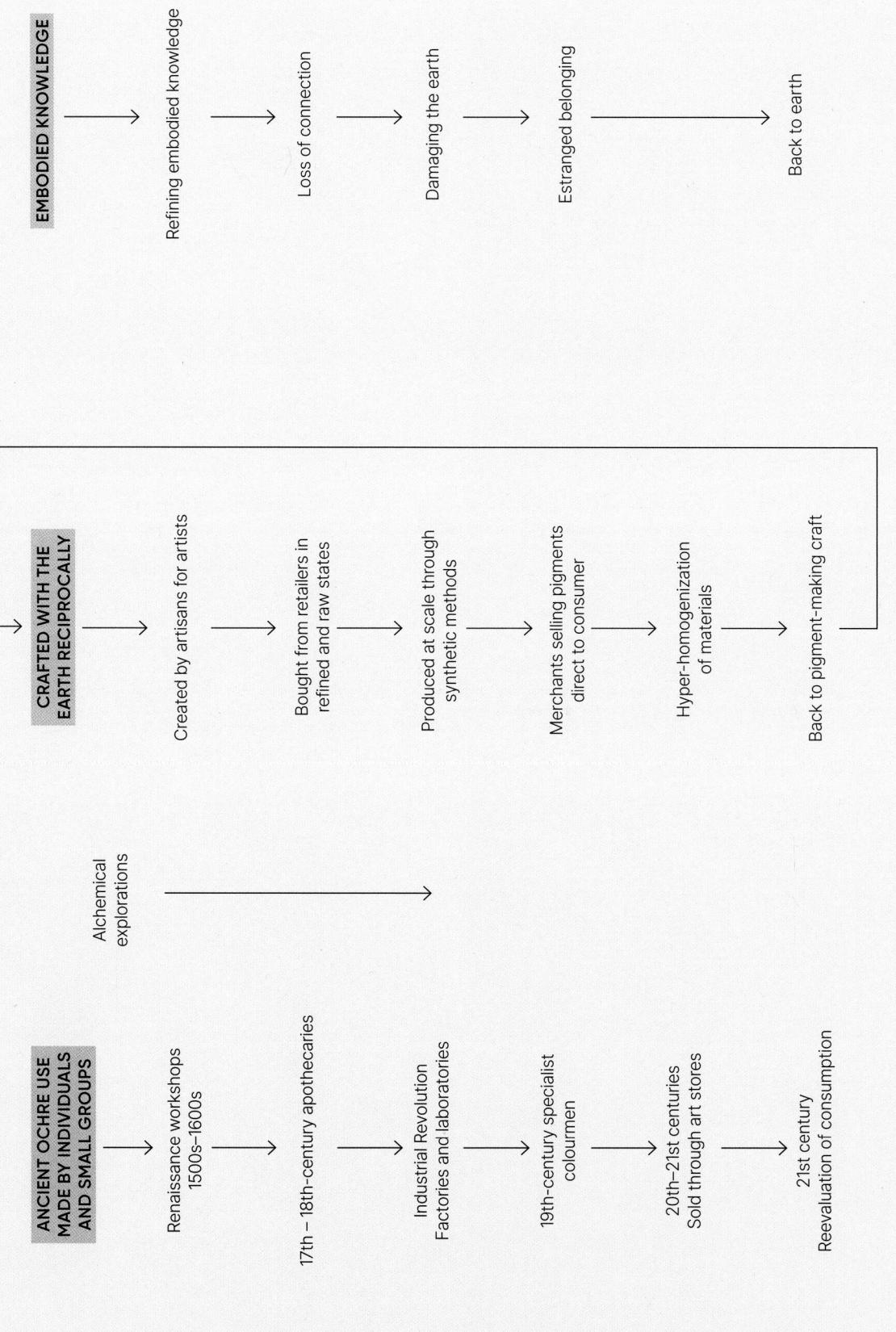

EFFECTS ON OUR INDIVIDUAL CONNECTION

EMBODIED KNOWLEDGE

Refining embodied knowledge

Loss of connection

Damaging the earth

Estranged belonging

Back to earth

STAGES OF PIGMENT MAKING EVOLUTION

CRAFTED WITH THE EARTH RECIPROCALLY

Created by artisans for artists

Bought from retailers in refined and raw states

Produced at scale through synthetic methods

Merchants selling pigments direct to consumer

Hyper-homogenization of materials

Back to pigment-making craft

Alchemical explorations

TIMELINE AND CONTEXT

ANCIENT OCHRE USE MADE BY INDIVIDUALS AND SMALL GROUPS

Renaissance workshops 1500s–1600s

17th – 18th-century apothecaries

Industrial Revolution Factories and laboratories

19th-century specialist colourmen

20th–21st centuries Sold through art stores

21st century Reevaluation of consumption

ORIGINS

Imagine the creation of a painting from the perspective of a single pigment particle – swished around and manipulated, lovingly jostled, tenderly nudged or energetically shoved into its final resting place, to be sealed in resin and oil, perhaps destined to endure for centuries. This pigment might be human-made or chemically refined, but often pigments are discovered, just as they are, in their natural state. They can be simply or lightly processed – cajoled into existence through arcane and ancient means: ground from shimmering minerals, coaxed from the essence of plants and animals, summoned from bones, collected from the earth's surface, or unearthed from its hidden depths.

The history of pigment starts with iron. Archeologists believe that coloured earth pigments have provided humankind with colours and paints for at least 500,000 years. Ochre pigments made from iron oxide, alongside other minerals in the same geological locality, were the first to be used. The use of pigments from the earth was a common and unifying activity of many cultures throughout the world, and they were one of the first materials utilized to externalize thoughts and emotions, aiding the development of the human brain.

As the fourth most-abundant element, iron makes up more than 85% of the mass of the earth's core and about 5% of the mass of the earth's crust. Iron readily reacts to water and oxygen, leading to oxidation, which forms iron-oxide compounds. When iron-bearing minerals weather over time, this leads to the accumulation of iron oxides and to the staining of other rocks by the iron. Water and rivers help the spread of iron throughout the land, contributing to the growth of ochre pigments over time. Iron is typically found in nature combined with other elements, and its ores and minerals are abundant and widely distributed.

Earth pigments are made of naturally occurring minerals, usually iron and manganese oxides. The dominant colouring matter in earth colours is iron, which comes in a variety of states and is generally accompanied by clays or quartz, alumina, carbonates,

micas, and metal sulfides. The exact make-up of earth pigments varies due to the geographic origin and the processing of the earth, and accounts for the different colours and their opacity or transparency when used in paint. Such mineral 'impurities' affect opacity and stability when used with a binder.

Ochre is the general name given to pigments that contain both naturally occurring and synthetic iron oxides, which are stable and usually non-toxic. The word derives from the Greek *ochros*, meaning 'yellowish'. The natural variations in the percentages of ochre's component parts is an important reason why the resulting pigments differ so much. Within one geological aspect, you might find a whole host of graduated colours, ranging from yellow, orange, and red through to green, violet, brown, grey, and black.

Ochre consists of three parts: the principle colour-producing component (hydrous or anhydrous iron oxide); the secondary or modifying colour component (e.g., manganese oxides within umbers or carbonaceous material within brown or black pigments); and the base or carrier of the colour (almost always clay, the weathered product of Aluminosilicate rocks). The yellow pigments contain iron- oxide hydroxides, while the reds are anhydrous iron oxides, and the darker colours can contain significant amounts of manganese. Today, the more transparent pigments are called 'siennas', the opaque ones 'ochres', and the darker ones 'umbers'.

THE COLOURS OF EARTH PIGMENTS

YELLOW OCHRE

Iron oxide hydroxide α-FeO(OH) with **clay minerals** (hydrous aluminium phyllosilicates) such as **kaolin** $Al_2Si_2O_5(OH)_4$

Soft, ochreous rocks containing hydrated iron oxide compounds known as iron oxide hydroxide are often termed yellow ochre. Limonite is the historical name for naturally occurring minerals containing yellow iron oxide hydroxides.

RED OCHRE

Anhydrous iron oxide Fe_2O_3 with **clay minerals** (hydrous aluminium phyllosilicates) such as **kaolin** $Al_2Si_2O_5(OH)_4$

Commonly found as a mixture of iron oxide with other clays or aluminosilicate materials, red ochres contain iron oxide in its anhydrous form, meaning it lacks water in its composition. Chemically similar to yellow ochre, red ochre typically forms through the weathering of hematite and other iron ores.

PURPLE OCHRE

Contains **iron oxide** (α-Fe_2O_3) and **manganese oxide** MnO_2 with accessory minerals such as **gypsum** $CaSO_4 \cdot 2H_2O$ and **calcite** $CaCO_3$

Purple ochre has a similar chemical structure to goethite but its different colour is due to variations in light diffraction properties, linked to a larger average particle size, as well as the presence of trace elements like manganese.

RAW UMBER

Contains **goethite** FeO(OH) and **manganese oxide** MnO_2 with accessory minerals **quartz** SiO_4, **kaolin** $Al_2Si_2O_5(OH)_4$, and **calcium carbonate** $CACO_3$

Raw umber pigments have a larger proportion of manganese oxide (5–20%), making them a dark brown; can be calcined to create burnt umber, where the goethite converts to hematite, to make a warmer shade.

RAW SIENNA

Contains **goethite** FeO(OH) and **manganese oxide** MnO_2 with accessory minerals **quartz** SiO_4, **kaolin** $Al_2Si_2O_5(OH)_4$, and **calcium carbonate** $CACO_3$

Raw sienna pigments have less than 5% goethite and only small amounts of manganese oxide, making them a rich orange-brown; can be calcined to create burnt sienna, where the goethite converts to hematite, to make a warmer shade.

GREEN EARTHS

Contains **glauconite** (K,Na)(Fe,Al,Mg) 2(Si,Al) 4O 10(OH) and **celadonite** K(Mg,Fe 2+)(Fe 3+ ,Al) [Si 4O 10](OH) 2 in different proportions, alongside accessory minerals **quartz** SiO_4, **kaolin** $Al_2Si_2O_5(OH)_4$, and the 37 **phyllosilicate mica minerals**

Glauconite – a type of mica often found in marine sedimentary rocks, e.g., sandstones and limestones – is a mineral made up of iron, potassium, aluminium, magnesium, silicon, and oxygen. Celadonite forms in arid contexts, in altered igneous rocks, such as basalts and andesites.

BLUE EARTH

Vivianite $Fe_2+_3(PO4)28HO$

The chemical name of this mineral is iron phosphate hydrate; it is responsible for medium bright blue earth pigments. The blue colour forms on exposure with oxygen, often transforming from a bright white to a deep blue over time.

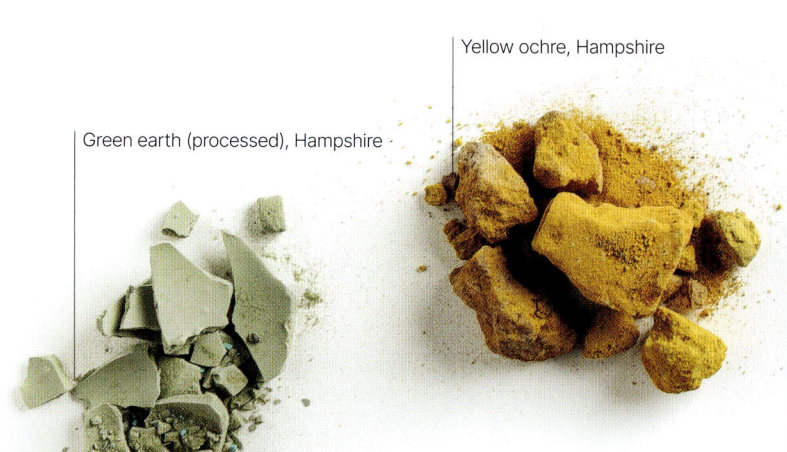

Yellow ochre, Hampshire

Green earth (processed), Hampshire

TRANSFIGURATIONS OF ANCIENT EARTH

Around 1.5 million years ago, humans harnessed fire for cooking, protection, and tool-making. In this section we delve into fire's pivotal role in altering earth pigments, the history of ochre, and experimental historical methods and recipes for enhancing and transforming the colours of natural earth materials.

The word 'calcination' is derived from 'calcine' (from the late Middle English *calcinen*), which in alchemical and medicinal parlance means to change the nature (of something) by heating; historically, it is related to the heating of lime for use in building. Early humans discovered that heating yellow ochre could transform it into red ochre. Similarly, when heated, natural sienna and umber pigments undergo dehydration, and the goethite within them is converted into hematite, resulting in the rich, warm tones that we recognize as burnt sienna and burnt umber.

Another ancient method is the levigation process, where mineral samples are graded and impurities, particularly sands and clays, removed by utilizing water. This technique has been used for thousands of years in various industries – including ceramics, pigment production, and cosmetics – to ensure a smooth, consistent product. In ancient Egypt and other early civilizations, it was commonly used to prepare fine pigments from natural minerals and earths. Artisans would levigate ochre or copper-containing minerals, such as malachite, azurite, and chrysocolla, to create high-quality pigments with a luminous clarity of colour.

Some of the earliest evidence of ochre processing, around 100,000 years ago, is in the Blombos Cave in South Africa, where early humans refined ochre for ritual and artistic use. Grinding stones reduced the particle size, and ochre pieces were intentionally shaped into drawing tools. (Similar techniques are found at Cueva de las Manos in Argentina, 13,000 years ago, and in the Lascaux Caves in France, 17,000 years ago.) Levigation purified ochre, and calcination enhanced its colour for art and decoration. These processes, used together, created a refined synergy in the extraction of both metals and pigments. Technologies that improved particle morphology were crucial to various industries, a role they continue to play today.

In Qafzeh Cave in Israel, archaeological excavations uncovered evidence of early *homo sapiens* using red ochre dated from 92,000–100,000 years ago. Ochre-stained tools and ochre-processing materials were found, along with 71 pieces of red ochre lumps. The people here collected and heated ochre to produce a more vivid red pigment in large hearths. This pigment was then used for various purposes, including burial rituals and potentially symbolic activities such as body painting. This discovery is among the earliest evidence of humans deliberately altering natural materials to create a pigment, marking an important development in the use of natural resources.

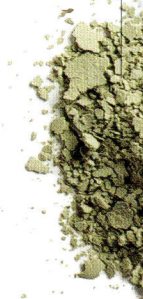

Green earth, yellow shade (processed), Hampshire

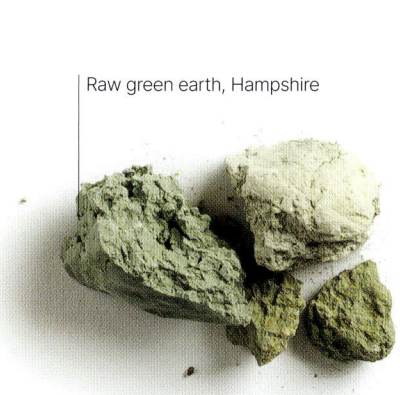

Raw green earth, Hampshire

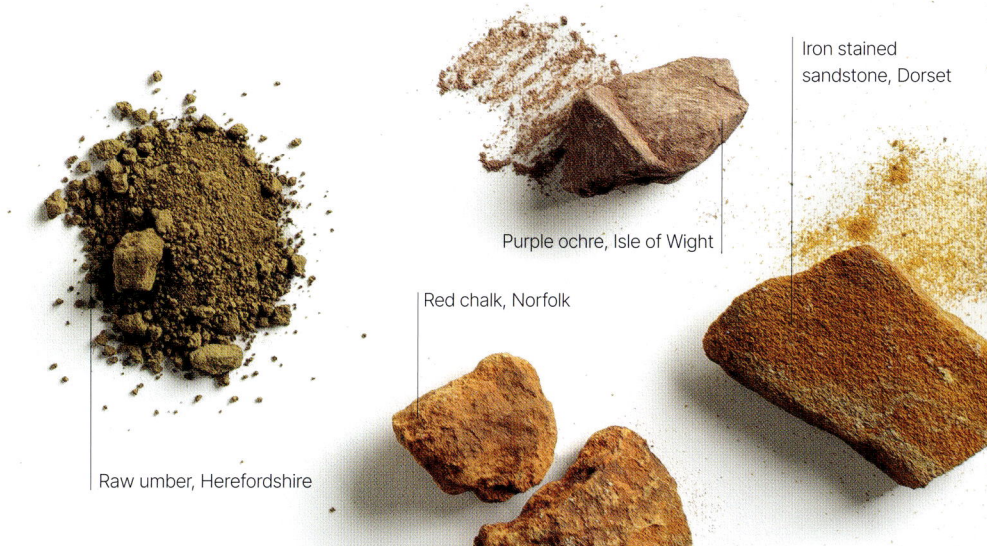

Raw umber, Herefordshire

Purple ochre, Isle of Wight

Red chalk, Norfolk

Iron stained sandstone, Dorset

RECIPE: CALCINATION

As changes to an ochre pigment can begin to happen as low as 100°C (212°F), there are some heating processes that are very accessible to those new to pigment making. For best results, use a levigated ochre sample that has the highest iron oxide content, or a purchased pigment such as synthetic yellow oxide or a bright yellow ochre. See Chapter 2: The Science of Pigment for instructions on making your own levigated pigments.

RECIPE 1: IN A PAN

EQUIPMENT

- Safety equipment (see box)
- Cast-iron or carbon steel pan
- Gas or electric stove
- Tongs or wooden scraper
- Jar for storing pigment

METHOD

Although temperatures are approximate and the highest temperature your pigment will reach will be about 370°C (698°F), heating ochre in a pan is an easy and accessible step into the world of calcination.

Pour finely ground ochre pigment into an old cast-iron or carbon steel pan (best for heat-retention); use approximately 50g in a 25-cm diameter pan. Heat on a gas or electric stove whilst stirring to evenly roast the pigment colour. (Alternatively, oven-roast on a tray for 30 minutes on highest setting.) Observe the pigment as the hue shifts; periodically remove some and set aside for colour comparison.

RECIPE 2: IN A KILN

EQUIPMENT

- Safety equipment (see box)
- Gas or electric kiln
- Crucible and kiln shelf or wire rack
- Tongs
- Jar for storing pigment

METHOD

Use a gas or an electric kiln, such as a small enamelling kiln (I recommend the Prometheus Pro) or an electric ceramic kiln.

Place about 20g of pigment in a crucible in the middle of the kiln. Be sure to use a wire rack or kiln shelf to protect your kiln from spillages. Choose even temperature intervals to create a selection of colours from the same sample. A good starting point would be increments of 100°C, 50°C, 20°C and 10°C (212°F, 122°F, 68°F and 50°F), but please follow heating instructions for your individual kiln. Individual samples to be calcined for 10 minutes each.

RECIPE 3: IN A FIRE

EQUIPMENT

- Safety equipment (see box)
- Dry wood, kindling, coal
- Metal biscuit tin
- Knife
- Tongs
- Jar for storing pigment

METHOD

Pierce several holes in the lid of your metal biscuit tin.

Place 50g of ochre pigment inside the tin and use tongs to place the tin in the middle of a roaring fire for several hours or until the fire dies down.

SAFETY EQUIPMENT

Wear heat-resistant gloves and safety eyeglasses when handling hot materials. Wear a dust mask or respirator and heat samples in a well-ventilated room.

Particulate protection: FFP3 Vapour protection: FFA1, FFB2P3, FFK1P2 (to European Safety Standards)

RECIPE 4: PIGMENT PRESS CAKE IN DIRECT FLAME

EQUIPMENT

- Saftey equipment (see box)
- Gas stove or fire or blowtorch
- Metal skewer or tongs
- Wire rack
- Jar for storing pigment

METHOD

Mix about 20g of your chosen ochre pigment with enough water to form a dry putty. Mix thoroughly and squeeze mixture around a metal skewer to form a pigment press cake (if it is too crumbly, add 10–20% kaolin to give it more strength and elasticity). Alternatively, roll the putty into lozenge shapes and use a pair of tongs.

Hold your pigment press cake over a direct flame until it changes colour. This may take around 10 minutes. By burning only one end you can compare the colours created. Once satisfied with the colour change, leave to cool on a wire rack.

If your shape holds together, try drawing with it; experiment with each end to see how the calcination has affected its colour and texture. If it falls apart, simply crush it using a pestle and mortar, and store pigment in a jar to use later.

TIP

Use a pestle and mortar to finely grind pigment, and use a 60 mesh sieve for a homogenous sample. Alternatively, experiment with larger lumps of ochre; however, be wary as thermal changes can make the samples crackle and explode.

AN ARTIST'S ASIDE

HONOURING INDIVIDUAL PIGMENT SOURCES

In the context of my own painting, I am reluctant to mix different pigments to get to a particular shade or colour combination. Labour and time is invested in coaxing a particular hue from its raw form into existence, so why would I want to adulterate a pigment that, in a way, is the condensed essence of a plant, rock or other starting material?

It is said that there was a tendency to use single pigment paints in the European medieval manuscript tradition, to keep the colours bright. This rationale was threefold: to stop painters from muddying their colours, to mitigate risk of the pigments chemically altering one another, and to appease superstitious anxieties surrounding the adulteration of them. Mixing pigments could be seen as an act that diminishes their individual qualities and a potential disservice to their original source, such as a plant or rock.

The fastidiousness needed to create a colour with chromatic purity or an unadulterated shade from raw materials requires some determination. Pigment makers inadvertently depend on nature's fluctuations, and many of my recipes rely on intuition rather than precise measurements. Rather than mixing and diluting a pigment to make a new colour, heat can be used instead. So it is fire that I choose to alter the shade, and even the hue, of my earth colours.

NOTABLE ARCHAEOLOGICAL SITES IN THE UK

Evidence of the use of ochre or iron oxide has been revealed at several sites, including:

Kents Cavern, Devon, 30,000 to 40,000 years ago: Excavations at this Upper Palaeolithic site uncovered evidence of ochre use, including ochre-stained tools. The site had been occupied by both Neanderthals and modern humans.

Paviland Cave (Goat's Hole), South Wales, 33,000 years ago: Situated on the Gower Peninsula, this limestone cave is famous for the discovery in 1823 of an Upper Palaeolithic burial of a young man, dubbed erroneously as the 'Red Lady of Paviland'. One of the earliest examples of ochre use in the UK, the bones were stained with red ochre, deliberately applied during the burial process, suggesting symbolic practices associated with death and possibly a belief in an afterlife.

Gough's Cave, Somerset, 12,000 years ago: Located in Cheddar Gorge, this cave is known for its evidence of human activity during the Late Upper Palaeolithic. Ochre-stained bones and tools were found, indicating the use of ochre in ritualistic or symbolic activities.

Star Carr, Yorkshire, 11,000 years ago: This Mesolithic site is one of the most famous for ochre artefacts in the UK, including more than 30 ochre-stained red-deer antler headdresses. These may have been used as a disguise when hunting, or during ritual performances by shamans when communicating with animal spirits.

Aveline's Hole, Somerset, 10,000 years ago: Late Upper Palaeolithic to Early Mesolithic burial cave, containing numerous human remains, some of which were stained with red ochre. One of the earliest cemeteries in the UK, it provides evidence of ochre use in funerary practices in the Mesolithic period.

INDIGENOUS USE OF PIGMENTS

Natural mineral pigments found in the earth, such as ochre, coloured clays, and malachite, have been used by Indigenous peoples across the world for millennia, serving not only as materials for artistic expression but also for ceremonial, cultural, and utilitarian purposes. Their connection to the land often translates into the use of these pigments in ways that reflect their environment, spirituality, and social structures.

For many Indigenous cultures, ochre is a sacred substance with profound spiritual, cultural, and symbolic significance. It represents the land, the ancestors, and the spiritual world, serving as a tool for healing, providing protection and connection to one's cultural identity.

Ochre is a foundational material for indigenous human activities, including rock art, sand painting, pottery decoration, ceremony, and face and body painting. These art forms are not just aesthetic but serve as vehicles for storytelling, cultural transmission, and spiritual expression. The use of ochre in art and ritual often involves the transmission of knowledge and techniques through generations, reinforcing the connection between the people, their land, and their heritage. In contrast, those of us who have not had such cultural and material knowledge entrusted to us need to explore ways in which art can be made from and with the land, while remaining respectful to Indigenous knowledge holders (ideas that will be explored in Chapter 6: The Future of Pigment Practice).

Note: I hope the inclusion of this section helps foster a deeper understanding of and connection with Indigenous peoples' histories and their futures. I strive to engage with and learn from Indigenous cultures in a sensitive manner; any generalizations made are not intended to cause offence.

AN INTRODUCTION TO MINERAL PIGMENTS USED BY INDIGENOUS CULTURES

INDIGENOUS PEOPLES	PIGMENTS	DESCRIPTION
Aboriginal Peoples of Australia	Ochre (iron oxide pigments)	Indigenous Australians have used ochre for at least 40,000 years, in rock art, body painting, and ceremony. Red, yellow, and white pigments were often ground from natural deposits and mixed with water or animal fat. Aboriginal rock art in regions such as Kakadu and Uluru in the Northern Territory includes depictions of animals, human figures, and sacred symbols. Ochre was also used to connect with the Dreamtime, the Aboriginal peoples' belief system related to the creation of the world.
Native American Tribes of North America	Red and yellow ochre, malachite, and hematite	Many Native American tribes, including the Plains Indians, used ochre and other mineral pigments in their artistic and ceremonial practices. Red ochre, made from hematite, was often used in burial rituals, and for body painting in both warfare and religious ceremonies. The Hopi and Navajo peoples utilized yellow ochre and malachite for sand painting – a sacred art form used in healing rituals. These pigments symbolized different elements of nature and spirituality.
Indigenous Peoples of Africa	Ochre, hematite, and kaolin	In Africa, ochre has been used by Indigenous groups such as the Himba of Namibia and the San (Bushmen) of southern Africa. The Himba women famously use red ochre mixed with butter to cover their skin and hair, a practice that is both protective and symbolic of their connection to the earth and ancestral spirits. The San, known for their rock art, used ochre and other mineral pigments to depict scenes of hunting, dancing, and spiritual ceremonies on cave walls.
Indigenous Peoples of South America	Red ochre, iron oxides, and other pigments	Many Indigenous groups across the Amazon, Andes, and Patagonia have used mineral pigments for body painting, pottery decoration, and ritual purposes. Among the Yanomami and Kayapo peoples of the Amazon, red ochre and other natural pigments are used in body painting for ceremonies and warfare. The Inca civilization used mineral pigments for wall paintings and textiles, often depicting religious iconography and scenes of daily life.
Indigenous Peoples of the Arctic	Red ochre and iron oxides	The Inuit, living in the Arctic regions of North America and Greenland, have historically used red ochre to paint tools, kayaks, and ceremonial objects. Ochre was also used in burial practices, where bodies were covered with the pigment to ensure safe passage to the afterlife. Inuit carvings and tools often featured pigments to enhance symbolic meaning or for functional protection against the elements.
Indigenous Peoples of South-east Asia and the Pacific Islands	Ochre and clay pigments	In Papua New Guinea and the Solomon Islands, mineral pigments are used extensively in body painting, especially during initiation rites, warfare, and tribal ceremonies. Red and white clays are commonly used to symbolize various spiritual meanings and status within the community. The Māori of New Zealand used ochre in tattooing practices, known as *Tā moko*, as well as in painting sacred meeting houses (*wharenui*).

PIGMENTS IN BRITISH FOLKLORE

The material nature of pigments, their properties and behaviours, are woven through the local traditions and ancient rites of British folklore. Natural earth pigments and plant-derived dyes often have mystical and medicinal properties, as well as practical uses in traditional crafts, and there is a deep connection between the earth, colour, and the supernatural.

The use of pigments for their transformative associations is captivating. By altering an object aesthetically, they may evoke or emphasize the protective power of the colourant, such as red ochre. This material holds a layered and intricate significance that is difficult to untangle – red ochre was cherished not only for its transformative qualities but also for its applications in medicine, magic, and protective rituals.

Natural pigments in British folklore have both specific/local and broad/global cultural associations. At Gradbach Hill in Derbyshire, for instance, near a now defunct ochre mine and processing site, locals claim that the 'Great Bitch' hill mother goddess was venerated. Located near The Roaches, a prominent rocky ridge in the Peak District, there are ochre settling ponds by the river Dane. A local deity was worshipped there for the medicinal earth that sprung forth from the waters. Nearby, ochres were added to sheep dips to protect their skin from flies and aid healing of surface wounds. Rams were painted with red ochre or 'reddle' to help identify those ewes that had been mated with.

These practices linked pigments to the earth and to concepts of life, death, and rebirth, drawing on the belief that certain colours and materials carried sacred or transformative power. Woad (*Isatis tinctoria*) is a blue dye made from the leaves of the woad plant, and it has a deep connection to early British history and folklore. The ancient Britons, notably the Picts, were once believed to paint themselves with woad before battle, as it was thought to imbue the wearer with courage and offer protection. This association has persisted in British legend, even though historical accuracy regarding its usage is debated, and may arise from woad's unique antimicrobial properties.

As we have seen, red ochre has been used since prehistoric times for cave paintings and body decoration. In British folklore, red ochre was often associated with protection and vitality due to its blood-like colour. In *Folklore of the British Isles* (1928) by Eleanor Hull, she notes, 'A mother should bring forth her child on the earth, or on straw, or she must lay her new-born babe on the earth, the source of life.' She also references ancient British superstitions surrounding ochre's potency, noting the belief that a person's death could be hastened if earth were placed on their chest – an effect said to be even stronger if they were laid on bare ground with naked feet.

There are ancient burial practices in parts of Britain where red ochre was sprinkled over the dead, symbolizing blood and the afterlife (the colour red had protective and transformative powers). In certain Celtic and pre-Celtic traditions, rituals involving red ochre were performed during seasonal transitions, especially around spring festivals like Beltane, symbolizing the earth's rebirth and the return of life. Red ochre appears to have held transformative properties when combined with water from springs and wells, particularly in connection with the festival of Imbolc, celebrated by Irish Celtic peoples and pagans.

Abundant in areas such as southern England, white chalk was used in folk rituals for protection. Chalk figures such as the Long Man of Wilmington (East Sussex) and the Uffington White Horse (Oxfordshire) utilize the chalk from the landscape to create monumental earthworks that serve as symbolic and cultural landmarks. Though their exact purpose remains somewhat unclear, these figures were likely

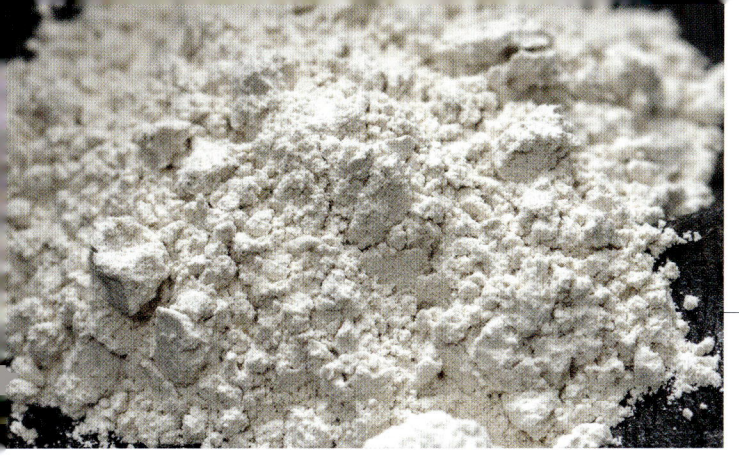

intended to mark territory, communicate religious ideas, commemorate events and be used in spiritual ceremonies, to protect against supernatural forces. Black pigments, meanwhile, made from charcoal or soot have long been associated with death, mourning and the afterlife. Black pigments were also used in magical workings to banish negative energy or curses. Carbon black paints were used for protective markings on houses. In some rural communities, black pigments were used in rites to ensure good harvests. Fields might be marked with soot or blackened wood as a form of sympathetic magic to promote fertility and ward off blight.

Many yellow and green pigments were derived from plants, such as dyer's greenweed, which are connected to fertility and renewal. As a reflection of the verdant greens of nature, these pigments offer a symbol of growth and renewal in Celtic traditions. However, this colour was also associated with fairies and the supernatural, leading to the belief that wearing too much green might attract unwanted attention from these otherworldly beings.

MEDIEVAL USES AND DISCOVERIES

There is evidence that pigments have been administered as drugs by physicians since ancient Greece. In Europe in the Middle Ages apothecaries were responsible for preparing and selling medicinal remedies, and often served as a bridge between medicine and herbalism. Apothecaries sold a wide variety of substances – including pigments such as ochre and verdigris – to be used in a variety of applications, from art and cosmetics to medical and alchemical purposes. Apothecaries supplied such pigments to artists, decorators and other craftspeople who used them in painting, manuscript illumination and other forms of artwork.

The word 'alchemy' is often floated about in reference to pigment making and the material changes that occur. This buzz word is used to describe fantastical transmutation, but it has a set of clear historical connotations. The history of alchemy – an ancient proto-scientific practice that combined philosophy, mysticism, and early chemistry – is deeply intertwined with the development of pigments. Its origins trace back to the ancient civilizations of Egypt, China, and Greece, where practitioners sought to understand and manipulate the materials of the natural world.

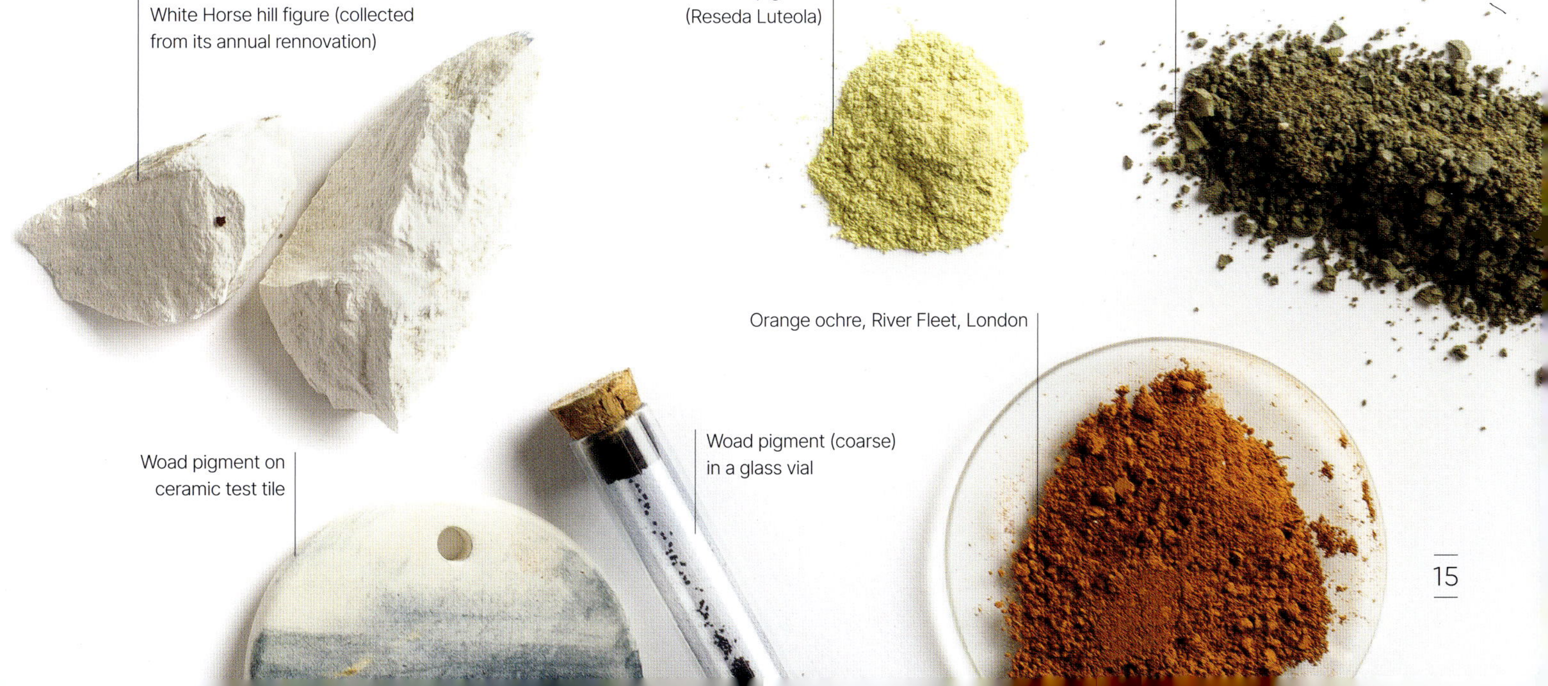

Raw chalk lumps from the Uffington White Horse hill figure (collected from its annual rennovation)

Yellow weld lake pigment (Reseda Luteola)

London Clay, Hackney

Orange ochre, River Fleet, London

Woad pigment on ceramic test tile

Woad pigment (coarse) in a glass vial

SYMBOLISM OF COLOURS

Alchemy, both a spiritual and proto-chemical tradition, is often described in stages that symbolize transformation – both of physical materials and the alchemist's own consciousness. These stages are traditionally associated with specific colors and pigments, each representing a phase of purification and transmutation. Though somewhat esoteric, this framework can offer valuable insight into both the outward transformations of pigment processing and the inner evolution of the maker engaged in the work.

BLACK (NIGREDO) Breaking Down of the Old	**Symbolism:** Destruction of impurities, ego death, and the beginning of transformation. **Process:** Calcination, burning, heating, or decomposition of a substance to break it down into ashes. **Pigments:** Carbon black, mineral black, manganese dioxide, lamp black.
WHITE (ALBEDO) Purification of the Essence	**Symbolism:** Signified purification and the washing away of impurities. **Process:** Extracting the pure from the impure, separating essential elements. **Pigments:** White lead (basic lead carbonate), chalk, gypsum.
GREEN (CONJUNCTION) Reuniting Opposites	**Symbolism:** The sacred union of opposites, balance, and rebirth. **Process:** Recombining purified substances to create something new. **Pigments:** Malachite, verdigris.
YELLOW (CITRINITAS/ FERMENTATION) Awakening and Enlightenment	**Symbolism:** Introducing new life or spiritual essence into the matter. **Process:** The infusion of divine wisdom, the solar principle, and the ripening of the work. Decay leading to new life, enlightenment, and inner illumination. **Pigments:** Orpiment, gamboge, aureolin, lead tin-yellow.
ORANGE-RED (DISTILLATION) Further Refinement	**Process:** Purification by repeated heating, evaporation, and condensation. **Symbolism:** Further elevation of the spirit and essence into pure form. **Pigments:** Minium, realgar, vermilion.
RED (RUBEDO/ COAGULATION) Perfection and the Philosopher's Stone	**Symbolism:** Completion, the elixir of life. **Process:** The final union of opposites, achieving enlightenment and transmutation into the highest form: gold. **Pigments:** Cinnabar, dragon's blood, red ochre.

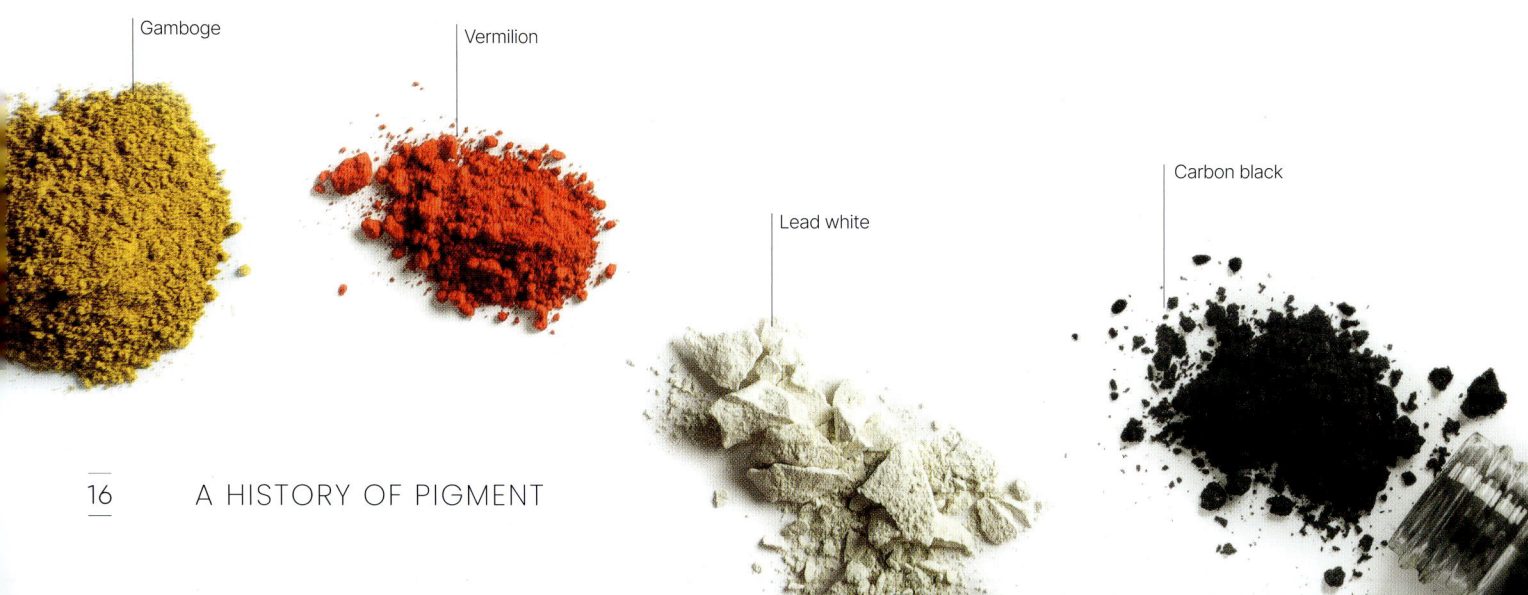

Gamboge

Vermilion

Lead white

Carbon black

Evolving through the Islamic world into medieval Europe, alchemy is most famously known for its quest to turn base metals into gold (the philosopher's stone) and the search for eternal life. Alchemists also made significant contributions to the discovery and refinement of pigments. They explored methods for extracting colours from minerals, plants, and metals, and sought to create new pigments through chemical reactions. These investigations laid the groundwork for the modern science of chemistry and the production of pigments.

One notable alchemical contribution to the world of pigments was the development of vermilion (mercury sulfide), a bright red pigment widely used in medieval and Renaissance art. Alchemists synthesized vermilion by combining mercury and sulfur through heating, demonstrating their understanding of chemical reactions and compounds.

In Renaissance Italy, artists such as Leonardo da Vinci and Sandro Botticelli were deeply interested in the alchemical concepts of transformation, and these ideas informed their art and their work with pigments. Artists and alchemists often worked side by side in workshops, sharing techniques for grinding and mixing pigments to provide themselves with their own unique palette of colours. Not all luminous ochres were to be found locally, however. A bright yellow ochre mined in England, at Shotover, Oxfordshire, was exported to Italy and was even used to paint parts of the Sistine Chapel.

Alchemy's influence on pigment production extended into the early modern period, as alchemical knowledge gradually transitioned into the field of chemistry. The experimental methods of alchemists laid the foundation for more systematic and scientific approaches to understanding matter, eventually leading to the chemical discoveries of the Industrial Revolution. Today, the rich history of alchemy can still be seen in the vibrant colours that were first explored in alchemists' laboratories, bridging the gap between mystical transformations and the practical needs of artists and artisans.

SYNTHETIC AND NATURAL PIGMENTS

The primary difference between natural and synthetic pigments lies in their origin and methods of production, yet both hold vital significance. Engaging with and studying both types allows pigment makers to deepen their understanding of colour, preserve cultural heritage, and open up new creative possibilities. While natural pigments are often made by hand using simple chemical and physical interventions under ambient conditions, synthetic pigments typically require controlled variables.

Synthetic pigments are produced by refining naturally occurring materials to isolate specific chemical compounds. This isolation enables the creation of purified substances and allows for the controlled exploration of their properties, particularly by examining the effects of specific impurities. In contrast, handmade pigment production operates along a continuum of material intervention. Without the conditions of controlled laboratory environments, trace impurities are an inherent feature of the final product. In handmade pigment production, there's a spectrum of intervention, from minimal interference with raw materials to more extensive processing and refinement. Without laboratory testing and tightly regulated processes, some impurities will always remain.

In the absence of controlled laboratory conditions, residual impurities are inevitable; yet it is often these very imperfections, such as minor mineral inclusions or subtle compositional variations, that confer natural pigments with their distinctive aesthetic qualities and enduring cultural significance. The study of synthetic analogues facilitates a better understanding of the role individual elements, such as sulfur in ultramarine and lapis lazuli play in chromatic and material behaviour. This analytical approach makes it possible to isolate variables that are otherwise difficult to study in natural specimens. Studying synthetic pigments helps reveal how specific elements contribute to a pigment's colour and behavior. In this way, many natural pigments are better understood through their synthetic counterparts.

Minium

Verdigris

KENDAL AND LINCOLN GREEN

Pigments have always embodied meaning and served as vessels of symbolism. They can evoke their origins, reveal the nature of their creation, or acquire significance through their application or use. Acting as conduits of communication, they derive significance from their relationships and context. The wielding of natural dyes such as woad and weld to create two lush greens is no exception.

Lincoln green is the name given to the colour of dyed woollen cloth originating from Lincoln, England, a major cloth-making town during the high Middle Ages. The cloth was well known across England and is often associated with high-quality textiles. It was created through a two-step dyeing process: first dyeing the cloth with weld (yellow), then overdying it with woad (blue). When combined, these two colours produced the distinct green shade that became associated with Lincoln.

Lincoln green is linked with legendary outlaw Robin Hood and his men, who were said to wear this colour. This association helped cement Lincoln green's place in English folklore as the colour of outlaws, hunters, and woodsmen, a camouflage for those living in forests or engaging in woodland activities. By the late medieval period, however, the production of Lincoln green began to decline, partly due to the rise of other cloth-making centres as well as changes in fashion.

Kendal green is associated with the Cumbrian town of Kendal in north-west England. The green-dyed, hard-wearing woollen cloth became especially popular in the late medieval and early modern periods. Closely associated with the working classes, it was often seen as a symbol of rural and military life and, in contrast to Lincoln green, it was firmly linked to everyday wear. Perhaps for this reason Shakespeare used this colour to help Falstaff describe a group of troublesome men *Henry IV Part I*: 'But, as the devil would have it, three misbegotten knaves in Kendal green came at my back and let drive at me.'

Kendal green was produced using a simple process that involved the direct dyeing of wool with a mix of natural dyes, most often woad and other green-producing dyes such as buckthorn or fern, to give a dark, subdued green suitable for utilitarian clothing. Production declined when more efficient textile-making centres led to cheaper mass-produced alternatives.

Below: Pictured here are lake pigments (using blends of woad, weld and verdigris) where I have converted soluble plant dyes into insoluble pigments. I see these shades of green as a colour to be worn by radicals; a usurper's anarchic green, a green to camouflage within and be radical under.

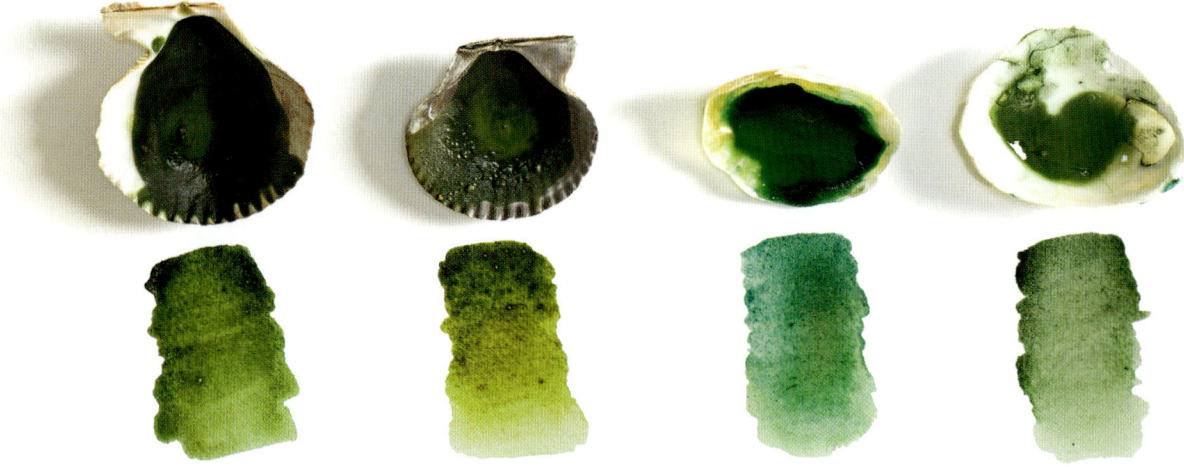

THE RISE OF SPECIALIST ART SUPPLIERS

As artistic techniques became more specialized, dedicated art supply shops began to emerge, particularly in cities such as Florence, Venice, and Antwerp. This shift from the provision of pigments by apothecaries to specialist art shops helped facilitate the growing professionalism of artists, and supported key art movements across Europe during the Renaissance and beyond.

Vendors known as 'colour sellers' (or *speziali* in Italy) supplied pigments, oils, and other materials specifically for artists. These vendors catered to painters, illuminators, and other artisans, offering high-quality pigments such as ochre, vermilion and ultramarine, along with materials for creating canvases, preparing paints, and mixing mediums. By the 17th and 18th centuries, dedicated art supply shops were common in all major cultural centres, serving the needs of amateur and professional artists.

In the UK, a similar evolution occurred. In London, for example, artists' colourmen – merchants who dealt specifically in artists' pigments and related materials – became increasingly prominent. One of the most

notable art supply shops was George Rowney & Co., established in 1783. Having initially operated as a stationers and apothecary, providing materials for artists alongside general goods, their focus shifted entirely to art supplies, and they became one of the most famous suppliers for artists in the 19th century.

This transition from apothecary to specialist art shop reflected the growing separation of the artistic profession from other trades, and the increasing demand for high-quality, reliable art materials. Artists no longer produced their own pigments, creating a disconnect from the materials and their origins. Instead, they came to rely on the services of art materials retailers.

Below: A selection of verdigris pigments and earth pigments (top right).

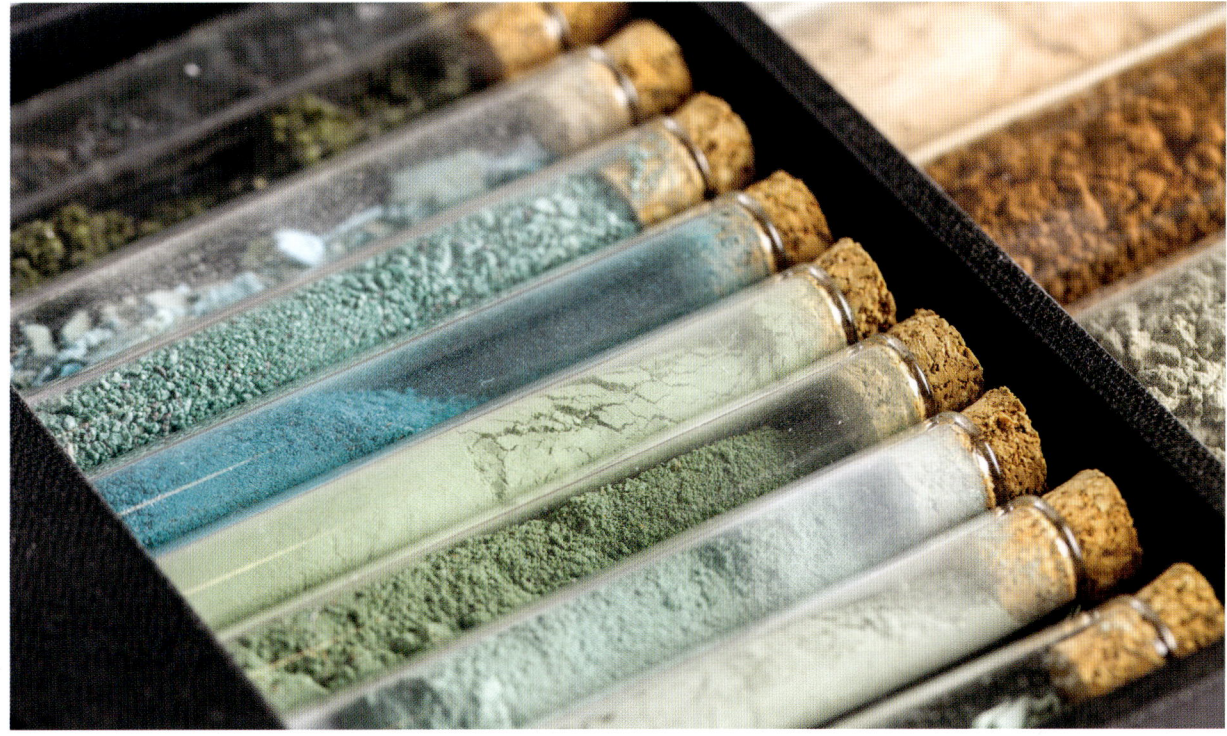

THE TRANSITION TO MODERN CHEMISTRY

The shift from alchemy to chemistry began during the Renaissance (14th to 17th centuries) and continued into the Scientific Revolution (16th to 18th centuries). Key figures such as Paracelsus, Robert Boyle and Antoine Lavoisier played pivotal roles in moving science away from the speculative goals of alchemy towards a more systematic, evidence-based approach fuelled by technological advances, new methodologies and shifting philosophical views. In this period, the alchemical processes of pigment making were refined into more reliable chemical methods.

The accidental discovery of Prussian blue, in 1704, was one of the first modern synthetic pigments, produced through chemical reactions involving iron salts and cyanide compounds. Not long after, synthetic red iron oxide pigments were being made in a laboratory setting. Called 'Mars pigments', they had all the properties, including durability and permanence, of their natural counterparts.

The production of synthetic pigments on a larger scale became a hallmark of the Industrial Revolution. In the early 19th century, the discovery of cadmium compounds led to the creation of cadmium yellow, orange, and red pigments. These vibrant, stable pigments became widely used in art. The techniques for isolating and purifying cadmium were direct descendants of alchemical metallurgy, as they required precise control of chemical reactions and heat.

Right: Raw materials for paint: Various ochres. Bottom image: gum prunus crystals. (cherry tree gum binder)

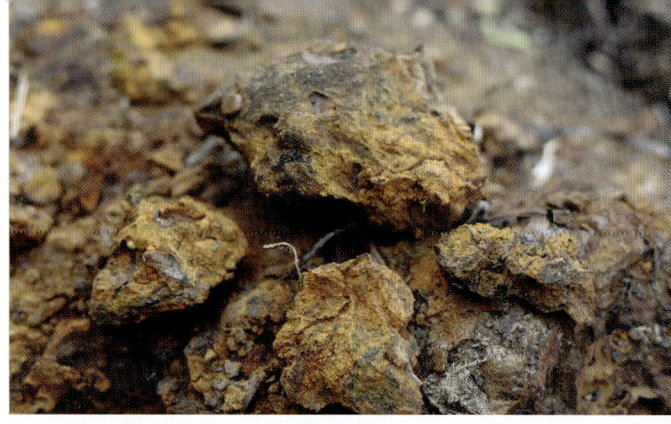
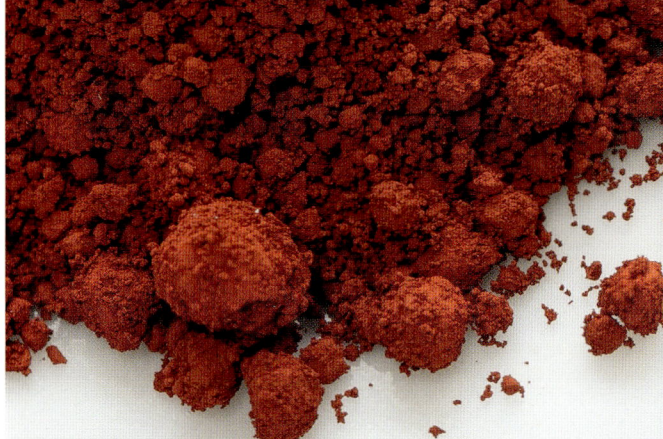
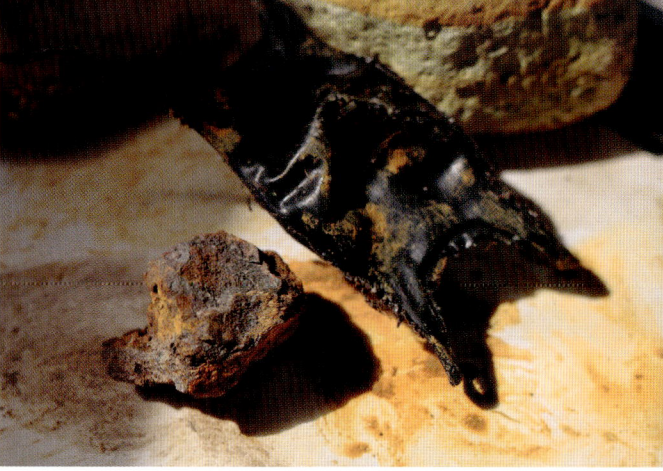
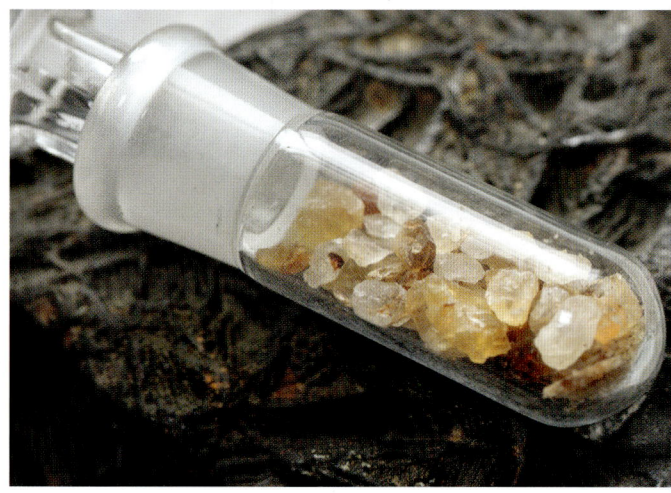

INDUSTRIAL INNOVATION

As we have seen, the history of pigment use spans millennia, with early humans using natural minerals and naturally derived substances to create colours for art, decoration, and symbolism. For centuries, artists were limited to a reduced palette derived from earth pigments, such as ochres, siennas, and umbers, as well as some minerals like lapis lazuli and malachite. These were often expensive and labour-intensive to produce, making brightly coloured art a luxury. The Industrial Revolution, however, transformed the landscape of pigment production and availability, introducing a wave of synthetic pigments that revolutionized art, manufacturing, and industry. The purest colours of the utmost clarity were now available to all. But at what cost?

In the late 18th century, significant advancements in chemistry altered the production of pigments. One of the most notable developments during this time was the accidental discovery of mauveine, the first synthetic dye, by English chemist William Henry Perkin in 1856. His vibrant purple dye led to the birth of the synthetic dye industry, making bright and durable colours accessible to artists and manufacturers, and opening the door to further innovations in chemical pigments. The significance of this development cannot be overstated, as it democratized colour, allowing for mass production. The chemical industry that grew from these early discoveries laid the foundation for modern colour chemistry, with profound impacts on both the visual arts and industrial design.

The first commercially available paints, in the form we recognize today, were introduced in the late 1700s. Before this time, artists typically had to grind pigments themselves and mix them with binders such as oil or egg yolk to create paint. This was time-consuming and required significant skill.

Modern watercolour pans, as we know them today, were invented by the English artist and entrepreneur William Reeves in 1781. He developed a method for producing small, portable cakes of watercolour paint that could be reactivated with water when needed, revolutionizing the way artists worked outdoors and while travelling.

The ingredients of his watercolour pans comprised a mixture of pigments (natural and synthetic, ground finely into a powder) combined with a binder, which was typically gum arabic. This natural gum, made from the sap of the acacia tree, held the pigment together and helped the paint to adhere to paper. Reeves' key innovation was the inclusion of honey or sugar-based syrup, which kept the watercolours semi-moist and allowed the paints to be easily dissolved with water and used for painting without the need for extensive preparation.

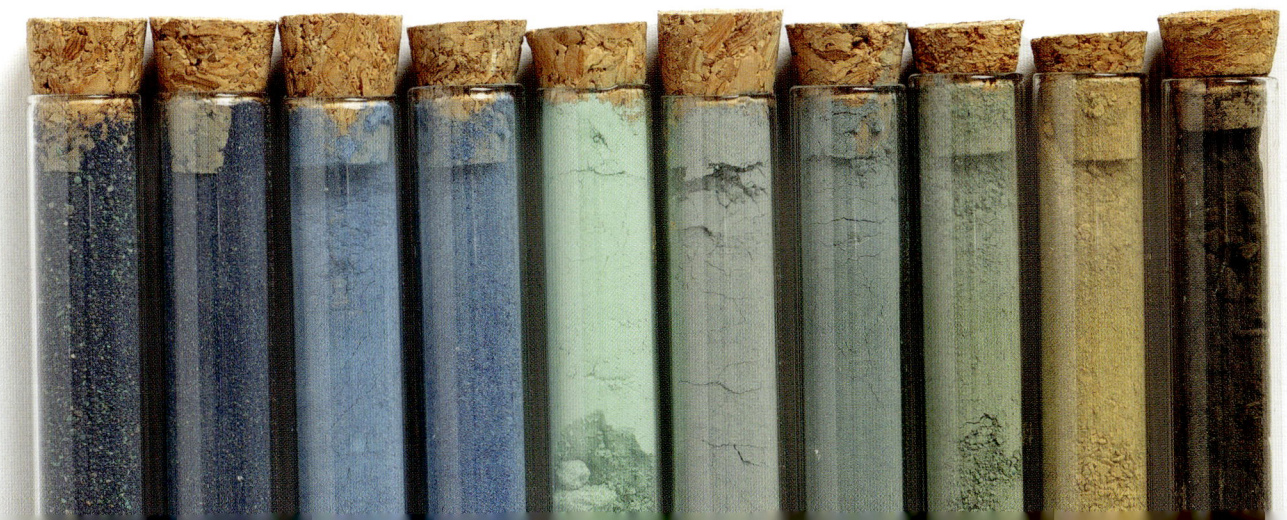

RECIPE: MAKE YOUR OWN WATERCOLOUR PAINT

Watercolour is one of the easiest paints to craft by hand, requiring only a water-soluble gum (e.g., gum arabic) and pigment. By adjusting the ratio of ingredients, you can customize your paint recipe. Remember, each pigment will need a slightly different amount of binder to achieve your preferred consistency. Grinding pigment to varying particle sizes provides valuable insight into how this can influence the behaviour of the paint.

INGREDIENTS

- Gum arabic in lump form
- Raw ground pigment
- Boiling water
- Honey
- Essential oil: clove, rosemary or lavender

EQUIPMENT

- Pestle and mortar
- Muslin cloth or fine sieve
- Grinding slab (polished granite or sandblasted glass) and muller
- Empty watercolour pans or half pans (or empty, clean seashells)

METHOD

1. Grind down a small quantity of gum arabic (about a tablespoon) using a pestle and mortar until it becomes a powder. (It is more economical to buy hard, brittle lumps than gum arabic in liquid form from an art shop.)

2. Dissolve 1 part gum arabic powder in 3 parts boiling water. Pour slowly and stir continuously for 10–15 minutes. (Alternatively, soak it overnight so that the gum absorbs the water, which will reduce the stirring time.)

3. Once dissolved, pour the mixture through a fine sieve or muslin cloth (to remove any bits of bark and other impurities that might have been contained in the solid gum arabic).

4. Add honey to the mixture – aim for 8 parts solution to 1 part honey, but be careful not to add too much, otherwise your paint will remain tacky after it dries. (Adding honey makes the paint more fluid and easy to work with later; without the honey, the pan will take a long time to get wet and 'release' any colour onto the brush.) Add 1–2 drops of essential oil for its antimicrobial effects.

5. Place a quantity of the prepared solution on the grinding slab, and mix in your chosen pigment, using a muller to grind it further. With an even pressure on the muller, grind the pigment in a figure of eight. Periodically scrape the mixture in from the side to incorporate. This will take 3–15 minutes depending on the pigment. It is important to achieve the correct balance of pigment to gum/honey mix: a general rule is slightly more gum than pigment. Some pigments require more than this, depending on particle structure. Each pigment, due to its own unique qualities, will require a different ratio of ingredients to create a homogenous mixture, so experimentation is key!

6. Pour the mixture into pans or half pans (or seashells) and leave them to set. If the pan cracks as it dries, make a note and next time add a little more gum solution to that particular pigment. Pour in thin layers to combat cracking as the paint is used, although this is merely for aesthetics.

Another key development that led to the commercialization of paint was the invention of the portable paint tube in 1841, by American portrait painter John G. Rand. He created a collapsible metal tube, similar to modern toothpaste tubes, which allowed paint to be stored for longer periods and in lightweight and convenient containers. Previously, pigs' bladders, tied with string or leather thongs, were used, and pierced with a nail to access the paint.

The paint tube revolutionized the art world, particularly for plein-air painters such as the Impressionists, who could now easily transport their paints and work outdoors more freely. Prior to this, commercially sold pigments and dry paint powders could be bought from colourmen. However, it was Rand's tube paint invention that made pre-mixed, ready-to-use oil or watercolour paint widely accessible to artists in a practical form. It led to an explosion of commercial paint brands in the mid-to-late-19th century, such as Winsor & Newton in England, which began producing artist-quality paints and quickly became a well-known brand. These developments greatly influenced the practices of artists, making the process of painting more efficient and allowing for more experimentation with materials.

Building on the significant advancements in chemistry of the previous century, the 19th century saw the creation of coal tar dyes, leading to the development of coal tar pigments. Key dyes such as mauve, fuchsine (magenta), and aniline yellow were chemically altered, making them stable for use in paints, inks, and coatings. Oxidizing agents such as chromium compounds, hydrogen peroxide, and potassium dichromate played a central role in this transformation.

These new synthetic colourants enabled vibrant, affordable colours to be used in mass production, including fabrics for the fashion industry, the automotive and decorative paints industries, and art supplies for professional painters. These pigments also found uses in the food industry and in the production of hair dyes, driving demand across a wide range of commercial products.

However, the introduction of coal tar dyes and pigments had significant environmental implications, both during their production and in their widespread use. Early coal tar products were often made using toxic chemicals and solvents, leading to hazardous waste and pollution. Manufacturing these synthetic dyes involved harsh chemicals, some of which were harmful to workers and the surrounding environment. Chromium compounds and other heavy metals used in the oxidation of dyes were particularly dangerous, contributing to soil and water contamination near industrial sites.

The widespread use of coal tar dyes also posed long-term environmental concerns. When they were eventually discarded, or washed off fabrics and other products, these dyes could leach into water sources, affecting aquatic life and ecosystems. Additionally, the petroleum-based origin of many synthetic colourants led to concerns about fossil fuel dependency and its associated environmental impact, contributing to the broader issue of industrial pollution during the era.

While advances in colour chemistry have led to safer, more environmentally friendly processes today, the early coal tar pigments were a significant environmental burden, laying the foundation for ongoing concerns about sustainable and non-toxic colourant production in modern industries.

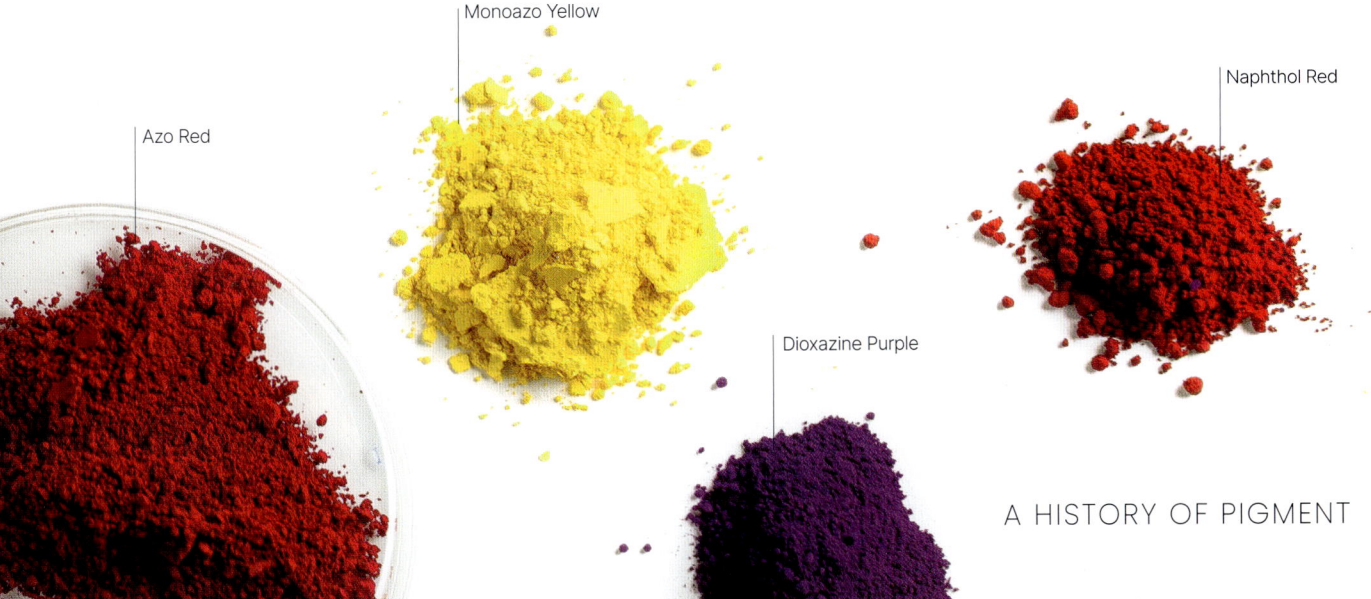

Azo Red

Monoazo Yellow

Dioxazine Purple

Naphthol Red

BACK TO EARTH IN THE 21ST CENTURY

In recent years, there has been a growing trend among artists to hand-make their own art materials, including paints, pigments, inks, and even papers. This movement can be seen as part of a broader return to traditional crafts and sustainable practices, where artists seek to reconnect with the origins of their materials and processes, often as a reaction against mass-produced, industrial products.

KEY ASPECTS OF THE HANDMADE ART MATERIALS MOVEMENT	
AUTHENTICITY AND CONTROL	Artists are increasingly valuing the ability to control the quality and properties of their materials. By hand-making their own paints or pigments, they can experiment with different textures, colours, and finishes that may not be available through commercial products. This process also allows for a deeper connection to the creative act, fostering a sense of authenticity in the final work.
SUSTAINABILITY	Many artists are concerned with the environmental impact of mass-produced art materials, which often contain harmful chemicals or are produced in ways that contribute to pollution. By making their own materials, they can opt for more sustainable, eco-friendly ingredients, such as natural earth pigments, or homemade binders using egg yolk, gum arabic or linseed oil.
REVIVAL OF TRADITIONAL TECHNIQUES	There is a renewed interest in traditional, often historical techniques of making art materials, such as grinding pigments by hand using a pestle and mortar or creating handmade paper. This revival is partly driven by a desire to preserve historical knowledge and craftsmanship, and partly by the unique aesthetic qualities that these methods can bring to contemporary art.
CUSTOMIZATION AND INNOVATION	Handmade materials allow for a high degree of customization. Artists can tailor their materials to specific projects, creating bespoke colours or textures that suit their artistic vision. Additionally, the process encourages innovation, as artists experiment with unconventional materials and methods, pushing the boundaries of what can be considered art supplies.
CONNECTION TO NATURE	Many artists involved in this movement emphasize a closer connection to nature by sourcing materials directly from their environment. This might include collecting natural pigments from the earth (such as ochres and clays) or using plant-based dyes and inks. This approach often carries with it a deeper philosophical or ecological message within the artwork itself.
COMMUNITY AND WORKSHOPS	The handmade art materials movement has also fostered a sense of community among artists. Workshops have become popular, both in-person and online, where artists share techniques for making pigments, paints or other materials. These workshops often emphasize collaboration, learning and the exchange of ideas, further strengthening the movement.

Craft practice is just that, an embodied potent repetition. Whilst practice may not make perfect, it slowly reveals its gifts. The brief historical outline of the evolution of pigment making and its use in creative practices, one could argue, has brought about a desaturation of ritualistic potency. We could see calcination, for example, as simply a way to create colours in controlled conditions, such as a kiln or a wood fire. But what we are really talking about here are the elements of fire, water, and earth – using fire we drive off water to reveal hidden properties of earth. Intuition comes to the fore in understanding the colour of the flame, the feel of the intensity of the heat, and the resulting colours.

By engaging with the materials they use differently, artists are re-engaging with traditional techniques, while innovating in ways that reflect contemporary concerns with sustainability, authenticity, and craftsmanship.

The move towards handmade art materials represents a shift towards a more thoughtful, personal, and sustainable approach to art-making. This movement is not only reshaping the art world but also creating a renewed appreciation for the process behind the materials, as much as the artwork itself. The re-evaluation and reinvention of urban and rural waste materials into high-quality artists' colours is one of the ways in which we can remediate our negative impacts on the planet in creative ways, that can transverse multiple industries and applications. This exponential potential for reconsidering the cultural hierarchies of materials is further explored in Chapter 6.

NATURAL PIGMENT COLLECTIVES

Social media platforms are filled with artists showcasing their hand-making processes, from sourcing raw pigments to crafting their own tools.

There are now many small-scale artisan brands that focus on producing handmade, artist-quality paints and materials, often using historic recipes and sustainable practices (see Useful Organizations and Pigment People for more advice).

AN ARTIST'S ASIDE

FINDING MY FOLKLORE

The folklore of pigments is often hard to trace due to the performative nature of oral traditions. I have been searching for my own folklore of pigments since I began using my own colours. I scoured libraries, online articles, and local museums, and interviewed pigment practitioners. But I was looking in the wrong places – I needed to look in myself. We tend to ignore the importance of our own experience and feelings, reverting to 'experts' and written proof. But what I've come to understand is that a deep artistic practice centred on handmade colour comes from within.

Below: Tools and paints made from natural materials

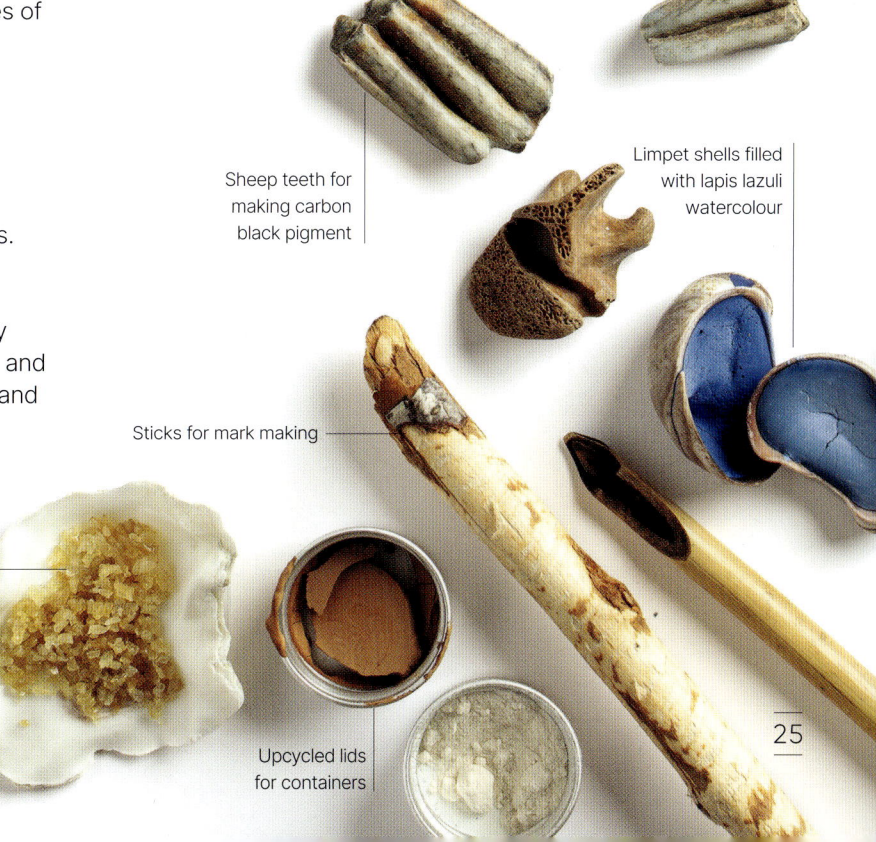

Sheep teeth for making carbon black pigment

Limpet shells filled with lapis lazuli watercolour

Sticks for mark making

Proteinaceous adhesive made from animal collagen

Upcycled lids for containers

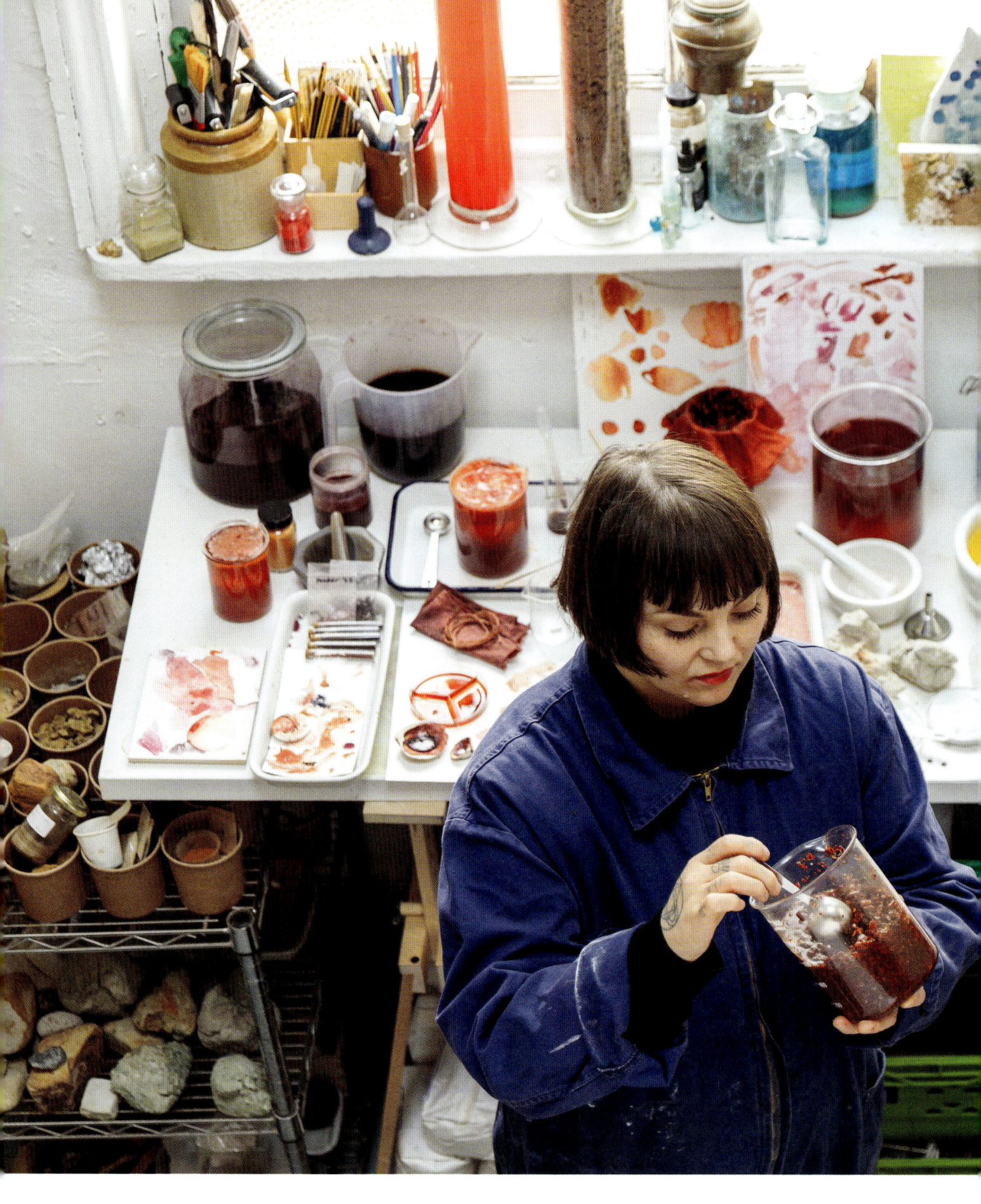

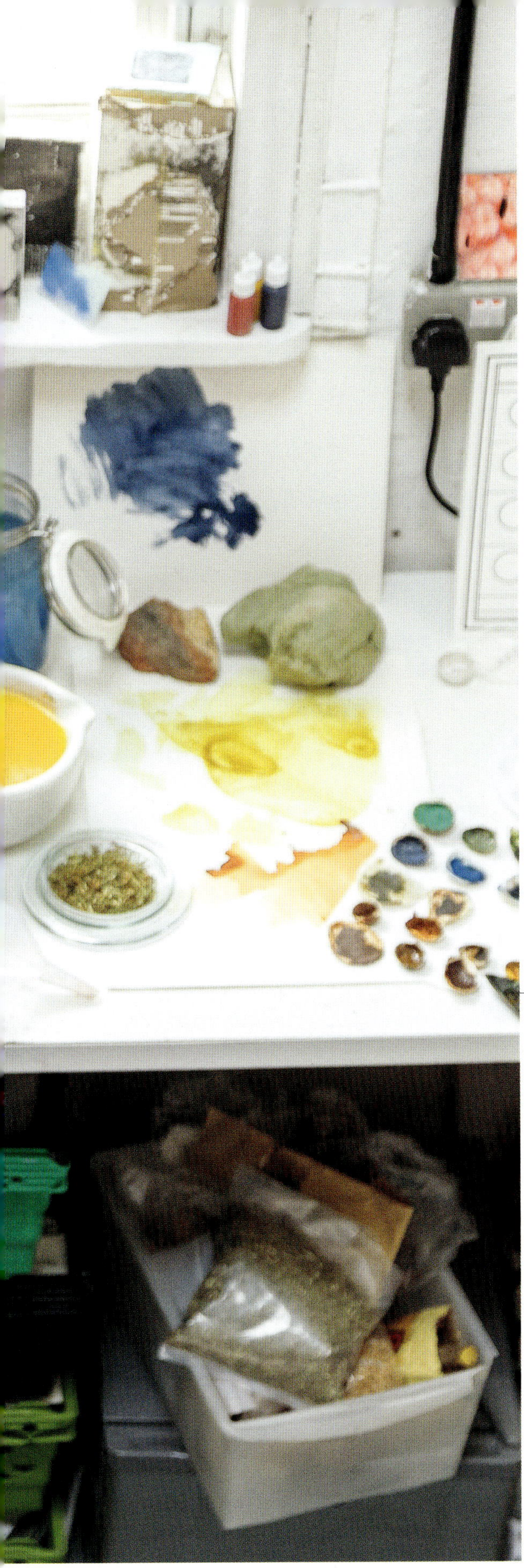

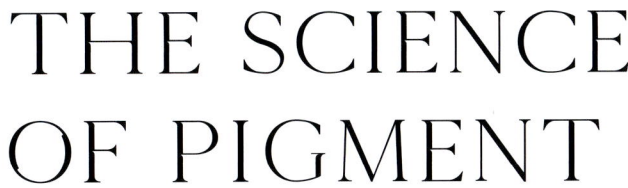

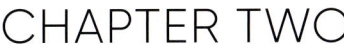

CHAPTER TWO

THE SCIENCE
OF PIGMENT

Behind every colour lies a mysterious
world of molecular interactions – chemical
and physical processes – along with the
interplay of light. These interactions often
include biological processes, either directly
or indirectly, with some pigments arising
from material from living organisms.

Understanding the science behind pigments reveals how abstracted colour has become in our society. It uncovers the specific ties that pigment processes have to places, materials, and local customs. The dissolution of these threads, once strong, has led to the emptying of authentic colour experiences and our deep, emotional connection to them. This chapter helps to unclothe some of the mysteries of their birth.

Some colours endure for centuries, both in use and in the fastness of their shade, while others fade and change. Some colours are toxic, while others are safe for use in art. The allure of natural pigments lies not only in their rich history and cultural significance but also in the fascinating chemistry that underpins their vibrant hues. By understanding the chemistry of pigments, we can better appreciate their role in art, industry, and everyday life, and equip ourselves with the technical knowledge to make informed colour choices in our work.

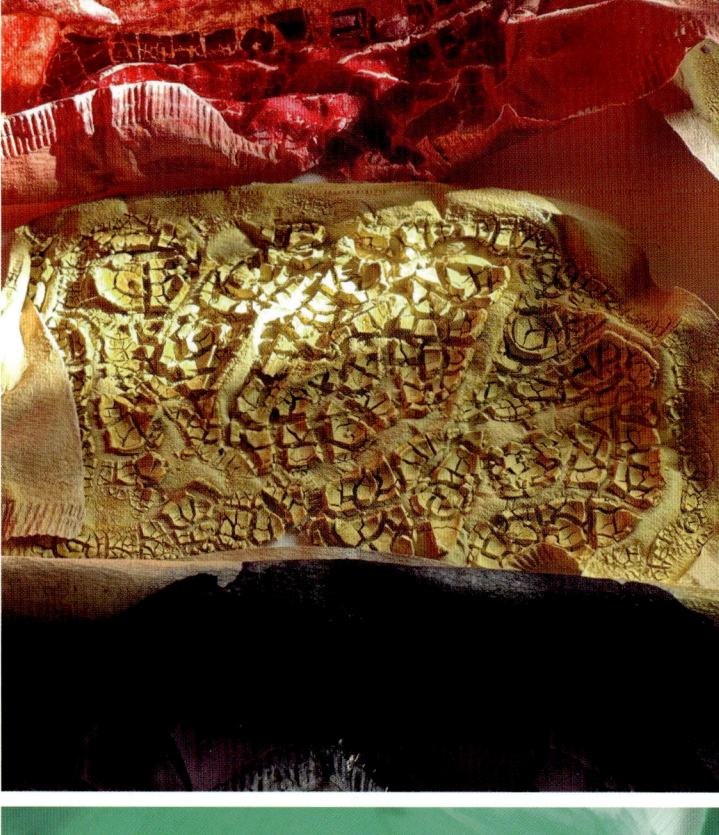

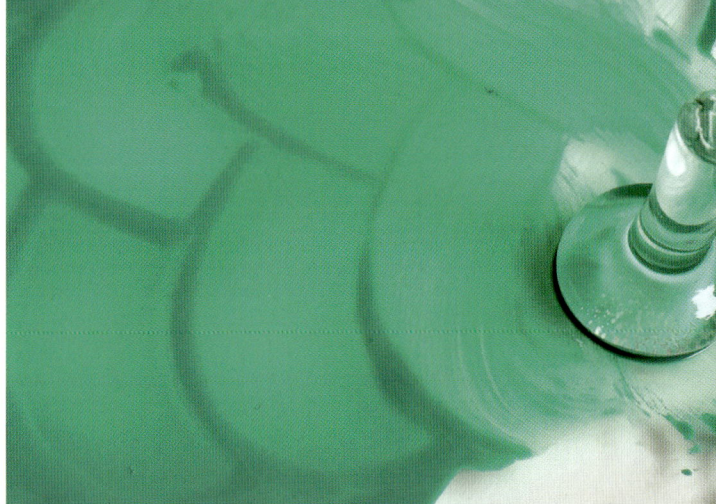

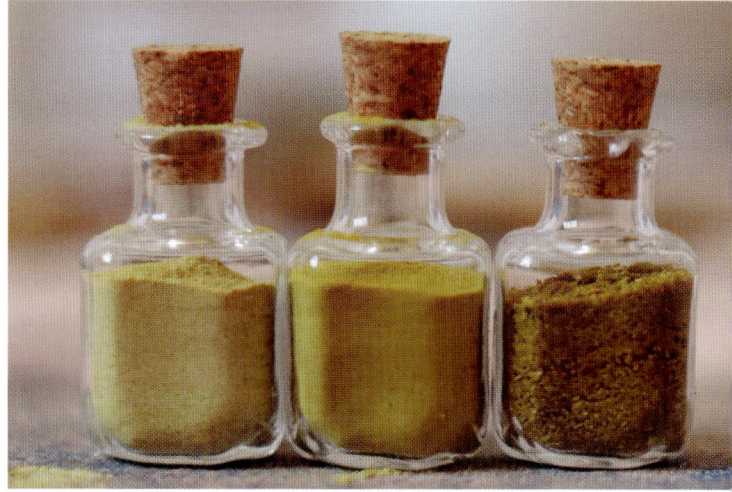

Right: (From top to bottom) Madder, weld and woad pigment drying out upon filter papers, verdigris watercolour paint being mixed with a glass muller & plate, three shades of ragwort (Senecio jacobaea) lake pigments.

WHAT ARE PIGMENTS?

At its core, a pigment is a colourant: a substance that imparts colour to other materials. This simple definition, however, belies the complexity of the interactions between pigments and light, as well as the intricate chemical structures that allow pigments to absorb, reflect, and transmit specific wavelengths of light.

Pigments are insoluble fine dusts, powders or particulates. They are distinct from dyes, which are typically soluble in water or other solvents. Dyes are fleeting and ephemeral – we call this 'fugitive' – as they slowly change over time, often desaturating into whispers of their former hues. This is because the connections between their atoms in their atomic structure are weaker and are broken down by ultraviolet (UV) light. Pigments are generally resistant to fading, and the ones used widely by artists or by industry are chosen for their inertness. Generalizations can beget inaccuracies. While pigments are usually insoluble and must be suspended in a medium (such as oil, water or polymer) to be applied to a surface, some are partially soluble in water, for example verdigris.

A pigment's colour is determined by its molecular structure – specifically the arrangement of electrons within its molecules. When light hits a pigment molecule, the energy from certain wavelengths of light can excite electrons in the molecule, causing them to jump from a lower energy level to a higher one. The energy required for this electronic transition corresponds to specific wavelengths of light. The wavelengths that are not absorbed are reflected back to our eyes, and these reflected wavelengths give the pigment its colour. For example, a pigment that absorbs red and blue wavelengths will reflect green, making the pigment appear green to the human eye.

AN ARTIST'S ASIDE

BE A PIGMENT MAKER – HAVE SOME GRIT!

'Grit' is my mantra, my guiding phrase, my repeated phrase of transformation.

Grit firstly speaks of the tangible world, of physical and mechanical processes. Within the discipline of geology it refers specifically to 'coarse sandstone', a sedimentary rock that consists of angular, sand-sized grains. Grit, dirt, and earth are useful tools for breaking down matter because of their fine (and often rather hard) particles. It is the fine dusts made from commonly found raw materials that I use to make pigment and paint. These dusts are then processed: washed, sieved, graded. (Discarded nodules of bricks, gravel or slate roof tiles have the potential to become dusts: fine urban grit.)

But its meaning is deeper than the corporeal or visible realm. To have 'grit' is to have courage and resolve, strength of character. You 'grit your teeth' when acts of sheer will and defiance are needed. Colloquially, if someone has 'grit' they have an 'edge' and shouldn't be messed with!

Grit is used to slowly break down larger masses, and the word is entwined with a sense of repetition and timescale. It is a reminder that paint, that infinitely malleable substance, is made of something quite different, derived from the earth. Fine grits are mixed with an adhesive liquid, such as linseed oil or gum arabic, to stick them down onto a surface in a fine coating or film. Making coloured dusts or pigment can be a hard and laborious activity – it requires a bit of grit in all its definitions to be a pigment maker.

CLASSIFICATION OF PIGMENTS

Pigments are classified as organic or inorganic based on their chemical composition, origins, and properties. This distinction helps differentiate between pigments derived from carbon-based compounds, often sourced from plants, animals, or synthetic organic chemistry, and those derived from minerals and metals. Organic pigments, known for their vibrant hues and translucency, contain carbon-hydrogen (C–H) bonds, while inorganic pigments, typically more opaque and durable, are made from metal oxides, sulfides, and other geological materials. Though some pigments blur these categories, such as organometallics, this classification remains essential in art, industry, and science, influencing colour stability, application, and production methods.

HISTORICAL BASIS FOR THE CLASSIFICATION OF PIGMENTS

The distinction between organic and inorganic chemistry stems from historical and structural differences rather than a strict natural divide. Initially, the 'vital force' theory held that organic compounds arose only from living organisms, but this was disproven in 1828 when Friedrich Wöhler synthesized urea from an inorganic source. Despite its origins, the distinction persists as a practical framework for studying materials, pharmaceuticals, and industrial chemistry.

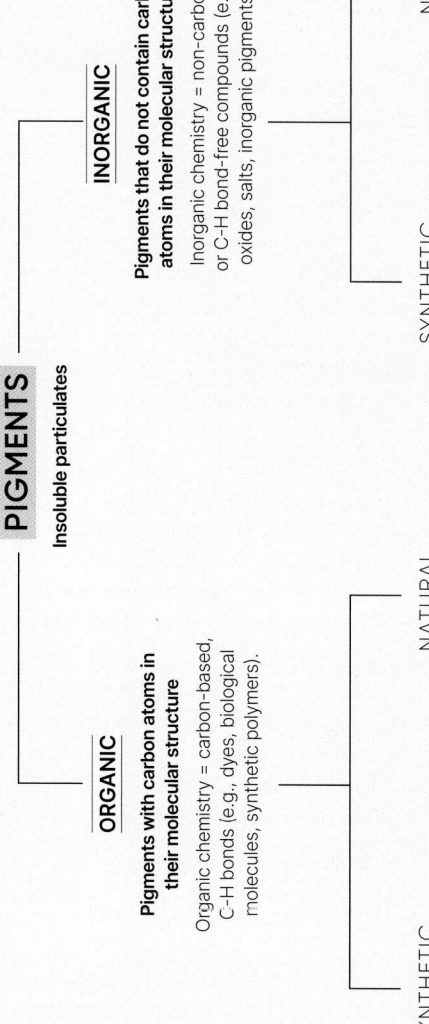

PIGMENTS
Insoluble particulates

ORGANIC

Pigments with carbon atoms in their molecular structure

Organic chemistry = carbon-based, C–H bonds (e.g., dyes, biological molecules, synthetic polymers).

SYNTHETIC

Definition: Human-made and highly processed pigments created from pure, high-grade chemicals.

Source: Derived from either **1** or **2**:

1. Chemically synthesized from petroleum-based compounds or coal tar derivatives, they contain carbon from once-living organisms.

Processed: Created through industrial synthesis, involving controlled chemical reactions to produce pure and stable pigments.

Examples: Alizarin, azo (yellow, orange and red colour range), phthalocyanine (blue and green colour range) and quinacridone (a lightfast red-violet pigment).

Colour Consistency: Highly consistent and pure.

2. Naturally occurring organic or synthetic dyes converted into lake pigments.

Processed: Extracted from raw materials, often requiring fermentation, boiling, or precipitation.

Examples: Cochineal lake, madder lake and weld lake.

Colour Consistency: Can vary due to natural impurities and environmental factors affecting plant/animal sources.

NATURAL

Definition: Naturally occurring pigments existing without human intervention.

Source: Materials that come from life: derived from plants, animals, fungi, bacteria and insects (e.g., madder root, cochineal, indigo).

Processed: Extracted from raw materials, often requiring fermentation, boiling, or precipitation.

Examples: Carbon black pigments, calcium carbonates.

Key Facts: Brightly coloured natural organic pigments without requiring artificial synthesis exist in their insoluble form but are rare to be found in nature: e.g., indigo, Tyrian purple, sepia (from cuttlefish ink).

Most exist with some human intervention, and as natural dyes converted into lake pigments.

Colour Consistency: Varies due to natural impurities.

INORGANIC

Pigments that do not contain carbon atoms in their molecular structure

Inorganic chemistry = non-carbon or C–H bond-free compounds (e.g., oxides, salts, inorganic pigments).

SYNTHETIC

Definition: Human-made, metal compounds.

Source: Derived from minerals, these are pigments produced through chemical synthesis rather than being naturally sourced.

Processed: Synthesized in controlled conditions.

Examples: Cadmium yellow/orange/red range, cobalt blue and titanium white.

Key Facts: The majority are brighter and last longer than their organic counterparts. Chemically identical inorganic compounds can occur in nature (e.g., lazurite in lapis lazuli), though they are rarely found in an uncombined state.

Colour Consistency: Highly consistent and pure.

NATURAL

Definition: Natural inorganic pigments that are mineral-based and occur naturally in the earth. They are used in their raw or minimally processed form.

Source: Geological materials such as minerals, clays, metal ores (e.g., oxides, sulfides, etc).

Processed: Mined, ground, and sometimes heated. Formed naturally through relatively simple chemical reactions, particularly oxidation. These metal compounds, such as oxides, are commonly referred to as earth colours.

Examples: Ochres (iron oxides) and ultramarine (derived from the mineral lapis lazuli).

Key Facts: Compared to organic pigments they are few in number.

Colour Consistency: Varies due to natural impurities.

THE SCIENCE OF PARTICLE CLUSTERING

In certain conditions, pigment particles clump together into larger, visible clusters within the paint. This is termed 'granulation'. In watercolour paints, it can create a textured, speckled effect, as heavier pigment particles settle into the crevices of the paper's surface. It occurs when the pigment particles are not uniformly suspended in the medium, leading to uneven deposition as the water evaporates. Whilst observing this phenomenon in your paint, consider the complex forces at play outlined here.

Attractive Forces: When the electrostatic repulsion between particles is weak, van der Waals forces and other attractive forces take over. These forces cause particles to cluster together loosely.

Brownian Motion: In a finely dispersed system, particles are constantly moving due to Brownian motion (random movement of particles in a fluid). However, when particle sizes increase due to granulation or flocculation, Brownian motion decreases, and particles are more likely to settle.

Flocculation: The process where pigment particles clump together to form flocculants or 'flocs' (loose clusters or aggregates). Pigment particles are colloidal in nature, meaning they are dispersed in a liquid medium but are small enough to be influenced by electrical charges on their surfaces. These charges typically repel one another, keeping the particles in suspension. When these charges are neutralized or reduced, the repulsive forces weaken, allowing the particles to come closer together and form flocs. Flocculation in paint can result in uneven colour distribution, with streaking or curdling effects where the pigment forms visible clusters within the paint film. This can be desirable in artistic contexts where texture and variation are appreciated, but it can also be problematic if uniform colour is required.

Hydrophobic/Hydrophilic Interactions: Some pigments have hydrophobic (water-repelling) surfaces, while others are hydrophilic (water-attracting). These surface properties affect how particles interact with each other and the medium. Hydrophobic particles are more likely to cluster together in an aqueous medium, leading to flocculation.

Liquid Medium Behaviour: The liquid medium plays a role in granulation. Water, for example, may not have sufficient surface tension to hold pigment particles in suspension, leading to their aggregation and the characteristic granular effect seen in watercolour paints.

Pigment Particle Size: Granulating pigments are usually composed of larger or more irregularly shaped particles. These particles do not disperse evenly in the liquid medium (e.g., water or oil). Instead, they tend to sink and cluster together due to gravity and van der Waals forces.

Polarity of Medium: The liquid medium can also affect flocculation. In water-based systems, changes in pH or the addition of salts or electrolytes can neutralize the charges on pigment particles, causing flocculation. Similarly, in non-polar solvents, the lack of significant electrostatic forces may lead to pigment aggregation.

Electrostatic Surface Charges: Pigment particles in a suspension often carry surface charges due to the nature of their chemical composition. These charges prevent the particles from coming too close to each other because of electrostatic repulsion. However, if these charges are neutralized (e.g., by changes in pH), the particles can aggregate, creating a granular texture.

MAKING LAKE PIGMENTS

The making of lake pigments involves the transformation of a soluble fugitive (quickly fading) colour into a new insoluble material that fades more slowly. They have been used throughout time for their bright chromatic purity. The term transitioned from the Latin word *lac*, meaning 'milk', reflecting the historical process of creating early lake pigments from dyes extracted from dripping resinous substances.

Lake pigments are created when a soluble dye is rendered insoluble by chemically binding it to an inert substrate (often a metallic salt or an inorganic base). This is typically achieved through processes like precipitation and neutralization, where the dye reacts with a mordant (such as aluminium, calcium, or iron salts) to form a stable, insoluble pigment. The colours created in this process are synthetic organic pigments derived from organic materials like plants and animals, or from synthetic dyes. These pigments are then artificially bound to a substrate, typically laboratory-grade alum, a processed and purified chemical. Ultimately, this procedure is akin to a conservation method that allows vibrant colours from organic dyes to be preserved and used for longer, coaxing out hidden colours within the source material which, without human intervention, we wouldn't be able to appreciate. As they are transparent, they are best used mixed with a binder (oil, resin or gum) and applied thinly in glazes.

From an artist's perspective, lake appears on the labels of art materials usually after a colour name (Violet Lake), or a scientific name referring to the specific compound or group of compounds (Quinacridone Lake), or an associative noun (Rose Madder Lake, Geranium Lake). Whether or not the name refers specifically to what is chemically in the tube, it will let artists know that the colour is vibrant, that it will be transparent or semi-transparent, and that it may fade over time.

AN ARTIST'S ASIDE

SUSTAINABILITY

Throughout the dyeing community, which uses the same chemicals for laking as for attaching dyes to mordanted cloth, there has been some urgency to make the procurement of chemicals more sustainable alongside water usage. It is possible to make some of them by hand, in the studio or the home.

Potassium carbonate can be made from taking hard wood ash, soaking it in water, filtering the plant material and evaporating the solution to reveal crystals of sodium carbonate (see Standard Laking Procedure for the recipe).

It is possible to forage for the substrate particle in some localities, where alkaline substances like alum, chalk, and limestone can be used in various recipes as the base material for attaching the dye molecule to.

Crucial in the lake pigment-making process are metallic salts, which allow water-soluble dyes to be converted into usable, insoluble pigments. Alum (potassium aluminium sulfate) is commonly used for red lake pigments (carmine, madder lake). Tin salts (tin chloride) are used to create primary red cochineal and madder pigments, which are highly toxic.

THE PURPOSE OF CONVERTING DYES INTO PIGMENTS

Colour Stability: Dyes are soluble in water or organic solvents, which makes them difficult to use in oil-based applications. By converting the dye into a lake pigment, the colour becomes insoluble, more stable, and less likely to react with the pH of the binder it is mixed with.

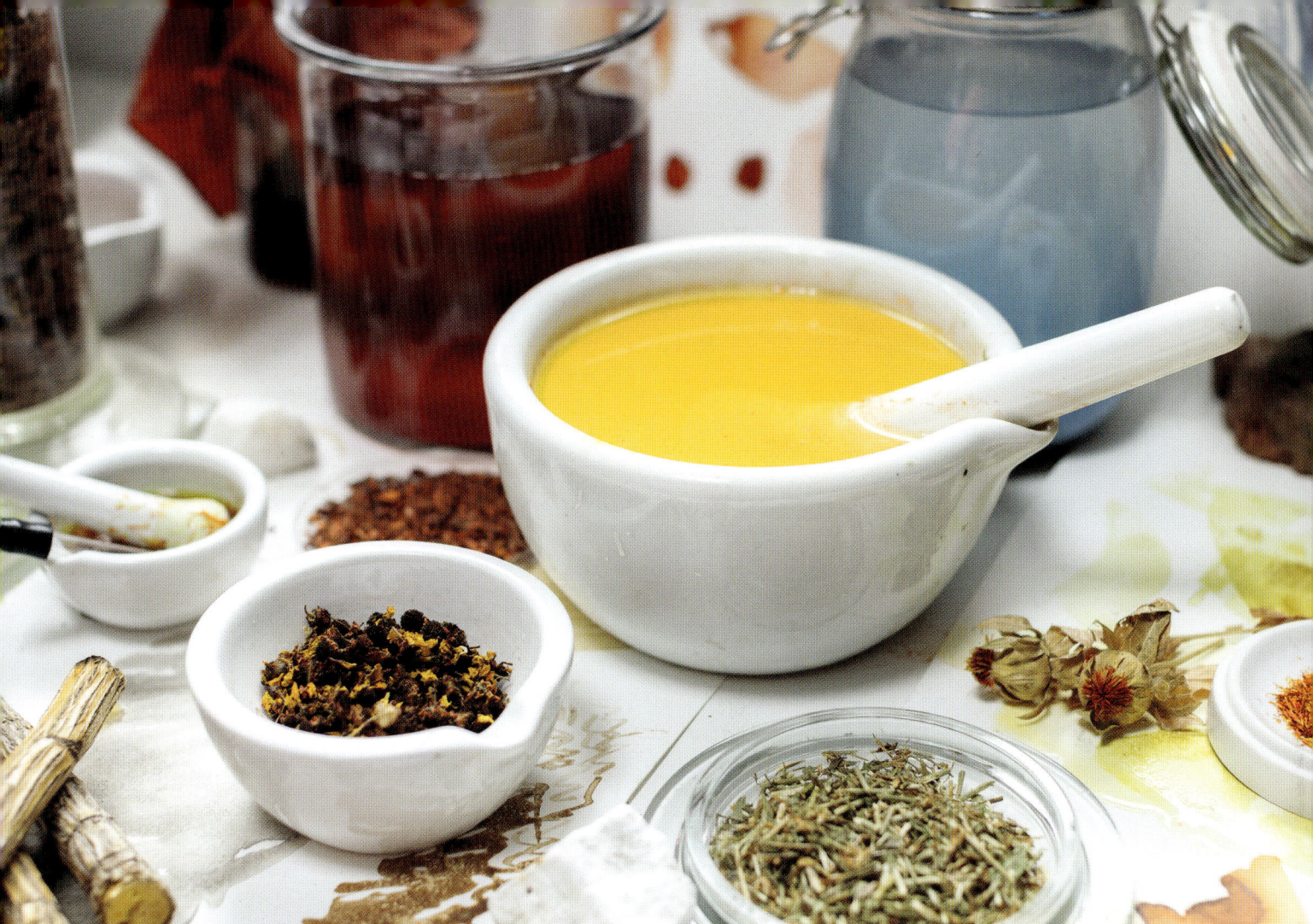

Vivid and Intense Colours: Lake pigments are known for their bright and often transparent hues, making them ideal for applications where strong, vibrant colours are desired, for art materials (watercolours and inks), textiles, and even food products.

Versatility: Because they are pigments rather than dyes, lake pigments can be used in a variety of mediums beyond just water-based solutions. They can be mixed into oils, waxes, and other materials without dissolving, making them highly versatile in manufacturing processes.

Improved Lightfastness: Many dyes are prone to fading when exposed to light. Lake pigments, being in a more stable, insoluble form, have better resistance to light, enhancing colour longevity in paints, cosmetics, and printing inks.

THE LAKE PIGMENT RECIPES

The lake pigment recipes in this book focus on the three primary colours, using madder and cochineal for red, weld for yellow, and woad for blue. These serve as the foundation for creating a wide spectrum of hues through admixture. The dyes featured here were classified as 'grand teints', originating from French for 'great dye', a designation reserved historically for high-quality dyes renowned for their exceptional durability and lightfastness, particularly during the peak of the natural dyeing industry in the early 1800s.

Recipes based on madder, weld, and woad yield vibrant and versatile colours with remarkable stability. When transformed into lake pigments, these dyes became some of the most enduring and reliable colourants from nature, epitomizing the height of botanical pigment-making artistry.

RECIPE: STANDARD LAKING PROCEDURE

A lake pigment is made by chemically attaching a dye molecule to an inert, substrate particle or structure to render it insoluble. The dye is chemically fixed in place by a metallic salt, usually potassium aluminium sulfate, commonly known as alum, and an alkali such as sodium or potassium carbonate is used to chemically 'fix' the two together. The metal ions in the salt react with the dye, causing it to bind to the solid substrate. This process forms an insoluble complex, allowing the organic dye to be used as a pigment in mediums such as oil and water-based paints. Making this general recipe is a good place to start learning about the process. You may wish to experiment with the ratios to optimize the pigment for your specific application. Use a dye extract from plant or animal material (dry or freshly gathered), but before removing plant material from a location, consider the ethical foraging guidelines first (see Chapter 6).

INGREDIENTS

- Dye source, 100g*
- Alum (potassium aluminium sulfate), 10g**
- Potassium carbonate, 5g***
- Distilled water

EQUIPMENT

- Fine muslin cloth; coffee filter papers and a funnel
- Glass vessels such as 500ml beakers
- Stainless-steel stirrers
- pH indicator strips

Recommended dye sources: weld, goldenrod, cochineal, brazilwood. If using fresh plant material use 50g instead of 100g.

**Use 10–20g of alum for every 100g of dried plant material.*

***See Recipe for Potassium Carbonate to make it yourself.*

METHOD

Note: record the exact proportions and quantities of your experiment as you work, for review in Step 8.

1. Extract your dye from your dry plant source (100g) by gently simmering it in water for 30 minutes with two fingers width of water above the plant material. Strain solution through a piece of muslin cloth.

2. Prepare a 10% alum solution by dissolving 10g of alum in 100ml of water. Heat may be used to dissolve the alum more easily.

3. Prepare a 5% potassium carbonate solution by dissolving 5g of potassium carbonate in 100ml of hot water.

4. Mix the dye solution with the alum solution. The ratio can be varied, but a starting point is to mix half up to equal volumes of the dye solution and the alum solution.

5. Slowly add the potassium carbonate solution to the dye-alum mixture while stirring continuously. During the reaction, carbon dioxide is given off so may result in a foam appearing on the surface of the solution. The reaction can be observed as the solution turns cloudy, and causes the formation of aluminium hydroxide, which acts as a substrate for the dye and forms the solid lake pigment. This solid material is known as a precipitate. Once no more visible reaction takes place, leave the solution to settle.

6. After the pigment has fully precipitated, filter the mixture through fine muslin cloth or a filter paper to collect the solid pigment. Wash the pigment with distilled water to remove any soluble salts that may still be present.

7. Allow the pigment to dry completely. You can spread it out on a flat surface or dry it in a low-temperature oven. (Once dry, it can be ground into a fine powder ready for use.)

8. Review your results: if your pigment is very light in colour, use more plant material in your dye extraction; too much alum in your ratio will result in it acting as a whitening agent in your final pigment.

RECIPE FOR POTASSIUM CARBONATE

You can make potassium carbonate – also known as potash – by hand, from locally sourced materials, and this can be used as a substitute for sodium carbonate in lake pigment recipes, simply swapping it out for the same weight.

1. Gather 10kg of hardwood (oak, maple, or beech), as these contain higher potassium levels; burn to make 1kg ash.

2. Place the ashes in a large container and add 5 litres of hot water. Stirring occasionally, let it sit for 12–24 hours to extract potassium compounds.

3. Pour the liquid (lye) through a fine cloth or coffee filter into another container, removing solid residues.

4. Boil down the filtered liquid in a heatproof dish until white potassium carbonate crystals begin to form.

5. Dissolve the crystals in 500ml of distilled water, filter again, then evaporate to obtain cleaner potassium carbonate (approx. 100g, depending on ash quality).

RECIPE: WOAD PIGMENT

Woad (*Isatis tinctoria*) is a hardy flowering plant that can grow in a range of climates and soils. Its striking blue colour played a central role in the dyeing textile industries of ancient Egypt, Mesopotamia, and various Celtic and Germanic tribes. During the Middle Ages, woad dyeing was an essential industry, particularly in England, France, and Germany. Woad was utilized as a blue colourant in European tapestry design from as early as the 1500s. These pigments were made by extracting the blue indigo-like dye from the woad leaves, which was then mixed with a binder to create a stable paint. With the advent of trade routes to the East in the 16th century, woad was gradually replaced by the more potent and cost-effective indigo dye. In the early 20th century, both were replaced by synthetic blue dyes. This recipe should yield approximately 2–4g of dry woad pigment, depending on the quality of the leaves and the efficiency of the extraction and precipitation processes.

INGREDIENTS

- Woad leaves, 500g (freshly harvested)*
- Sodium carbonate (soda ash), 20g
- Distilled water
- White vinegar, 100ml

EQUIPMENT

- Fine sieve or muslin cloth
- Coffee filter papers and a funnel
- Large inert cooking pot (glass or stainless steel)
- Wooden spoon/stirrer
- Pestle and mortar
- Dehydrator (optional)

It's important to use freshly harvested leaves;(ready for picking in the UK in July and August).

METHOD

EXTRACTING THE DYE

1. Wash the woad leaves to remove dirt or debris, and finely chop to facilitate dye extraction.

2. Place leaves into the pot and pour over 2 litres of hot water (approx. 80°C/ 174°F – just below boiling to avoid degrading the dye). Steep for 10–15 minutes while stirring gently to extract the maximum amount of dye.

3. Strain the leaves out of the solution using a fine sieve or muslin cloth, pressing out as much liquid as possible.

ALKALIZING THE SOLUTION

1. Add 20g of sodium carbonate and stir thoroughly to ensure the alkali is fully dissolved, the pH should be between 9 and 10. The solution will turn a greenish colour, indicating that the indigo precursor molecules in the woad have been released.

PRECIPITATING THE PIGMENT

1. Begin to stir the solution vigorously to introduce air. This oxidation step will cause the reduced indigo to revert to its blue, insoluble form, precipitating out of the solution.

FILTERING AND WASHING

1. Allow the blue pigment to form and settle, then carefully pour the liquid through a filter paper placed in a funnel to collect the indigo particles, which appear as a paste or small clumps.

2. Use distilled water to rinse the collected pigment through the filter to remove any residual alkali or impurities, repeating until the water runs clear.

3. Wash the pigment with 100ml of vinegar diluted in water, to neutralize any remaining alkaline residue; then wash in distilled water to remove the vinegar.

DRYING AND GRINDING

1. Spread the wet pigment on a flat surface and allow to air dry in a warm, dry place (or use a dehydrator).

2. Once dry, the pigment can be ground into a fine powder, using a pestle and mortar. Store in a dry place out of direct sunlight.

PIGMENT SUMMARY DEFINITIONS

DENSITY Refers to the mass of a pigment relative to its volume, influencing how it settles in a binder or suspension. It is measured in grams per cubic centimetre (g/cm^3) by dividing mass by volume.

HARDNESS Indicates a pigment's resistance to scratching, which affects how easily it can be ground into a fine powder. Measured by the Mohs scale from 1 (talc) to 10 (diamond), which tests whether a material can scratch or be scratched by reference minerals.

REFRACTIVE INDEX A measure of how light bends as it passes through a pigment, affecting its opacity and brilliance. Pigments with a high refractive index (around 2.70) appear more opaque, while those with a lower index can be more translucent. It is calculated based on the ratio of the speed of light through the pigment compared to air.

PIGMENT SUMMARY

Pigment Name: Woad (*Isatis tinctoria*)

Nomenclature: Dyer's woad, glastum, Asp of Jerusalem

Pigment Index Number: NB 1

Colours: Blue, green

Colourant & Chemical Formula: Indigotin $C_{16}H_{10}N_2O_2$ (dye precursor indican, a glycoside that ferments to become indoxyl and then oxidizes to indigotin)

Classification: Synthetic organic

Description: Indigotin precipitates as an insoluble solid after indoxyl oxidizes after extraction

Conservation Issues: Will fade if subjected to direct sunlight

Particle Morphology: Crystalline, angular, elongated or needle-like in shape. Fine and small, likely to aggregate; can have a rough and irregular surface and shape.

Dates in Use: Neolithic period (Europe) to 19th century

Density: 1.35–1.40g/cm³

Hardness: 1–2 on Mohs scale

Oil Absorption Ratio: 40–60g of oil per 100g of pigment

Refractive Index: 1.66 (moderate to decent opacity and light-scattering properties)

Health and Safety: Non-toxic

Environmental Impact: Woad has minimal environmental impact when grown sustainably. However, intensive farming or non-ecological farming practices can lead to soil degradation or pesticide use. See also Environmental Impact of Laking Chemicals.

RECIPE: WELD LAKE PIGMENT

Weld (*Reseda luteola*) is a biennial, growing up to 1.5m (5ft) in height. It has long spikes with small pale yellow flowers that start appearing in early June. One of the oldest and most significant sources of yellow dye, weld was used to create bright yellow colours in textiles and pigments, and prized for its brilliance. Said to have been used to dye the robes of the Roman vestal virgins in ancient Rome, in medieval times it became a staple in the European dyeing industry. It was also used in conjunction with other dyes, such as indigo, to produce vibrant greens. When over-dyed with woad, it produces Lincoln green (see Chapter 1).

The weld lake pigment, made by precipitating the yellow dye onto a stable substrate like aluminium hydroxide (alumina), was widely used in painting, including manuscript illumination. In the 19th century, chemists refined the production of lake pigments with weld, making it a stable and long-lasting yellow pigment. It can be mixed with gum arabic for use in watercolour painting or with oil for oil painting. It produces a bright, transparent yellow with excellent lightfastness, making it a valuable addition to any artist's palette.

This recipe follows traditional methods while adapting them for modern use. It ensures that the resulting pigment is stable and suitable for use in a variety of artistic media. It should yield approximately 10–15g of dry weld lake pigment, depending on the efficiency of the extraction and precipitation processes.

INGREDIENTS

- Weld leaves or flowering tops, 200g, fresh or dried*
- Alum (potassium aluminium sulfate), 20g
- Sodium carbonate (soda ash), 10g
- Distilled water, 1.5 litres

EQUIPMENT

- Fine sieve or muslin cloth; coffee filter papers
- Large inert cooking pot (glass or stainless steel)
- Glass or plastic containers
- Pestle and mortar

Include seeds for strongest dye; note that the woodier parts, such as the stalks, contain tannins to create mustard shades.

METHOD

EXTRACTING THE DYE

1. If using fresh weld, chop the plant material into small pieces to facilitate the extraction of the dye.

2. To ferment your plant material, leave it soaking overnight with enough water to cover it. (The plant material can be weighed down if necessary.) It can be left soaking for up to four days. See Fermentation for more information.

3. Place the weld plant material in the pot and add 1 litre of distilled water. Bring the water to a gentle boil and allow to simmer for 30–60 minutes, stirring occasionally. The water will take on a brown-yellow hue as the luteolin is extracted from the plant.

4. After simmering, strain the weld solution through a fine sieve or muslin cloth to remove the plant material. Set the yellow liquid (the weld dye solution) aside.

PRECIPITATING THE PIGMENT

1. In a separate container, dissolve 20g of alum in 250ml of hot distilled water. Stir until fully dissolved.

2. Slowly pour the alum solution into the strained weld dye solution while stirring continuously. The alum will react with the luteolin in the weld, beginning the process of forming the lake pigment. Typically, the solution will look more opaque with more of a yellow hue.

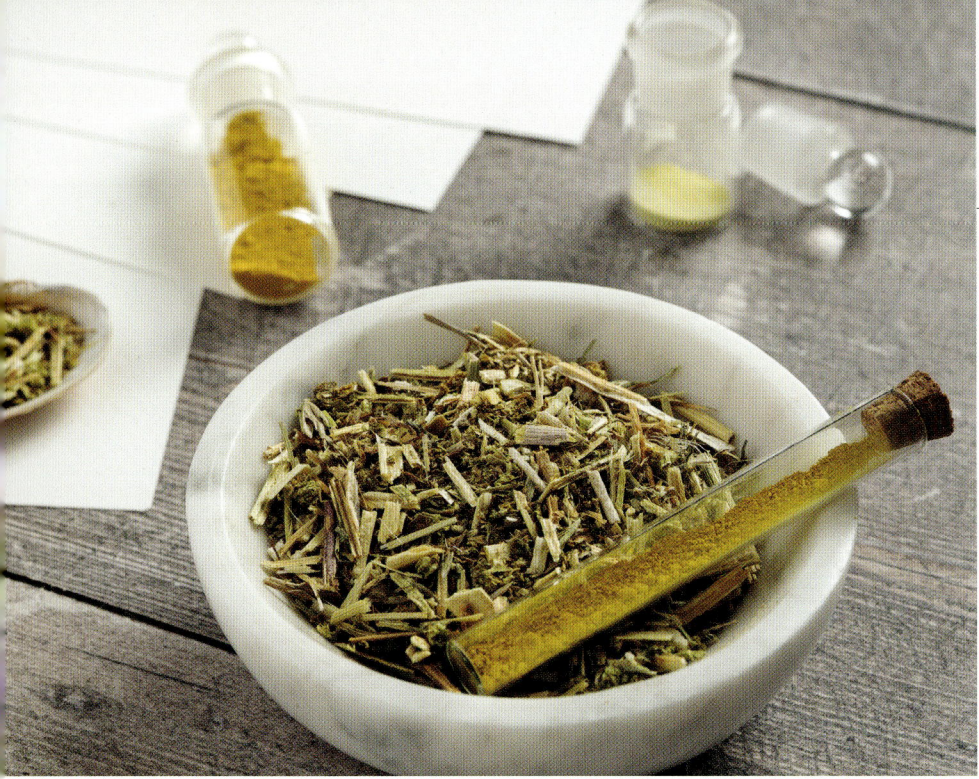

PIGMENT SUMMARY

Pigment Name: Weld (*Reseda luteola*)

Nomenclature: Dyer's greenweed, dyer's weed, dyer's rocket, woold, dyer's mignonette

Pigment Index Number: NY 2

Colours: Bright lemon yellow, greenish yellow, mustard yellow

Colourant & Chemical Formula: Luteolin (flavonoid) $C_{15}H_{10}O_6$

Classification: Synthetic organic

Description: Most lightfast of the yellow lake when mordanted with alum

Conservation Issues: Fugitive if in direct sunlight, may fade over time if subjected to extremes in pH, high humidity or highly polluted environments

Particle Morphology: Amorphous hydrated alumina, fine grained, irregular, highly dispersible pigment

Dates in Use: Significant in medieval and Renaissance Europe; up to 19th century when synthetic dyes became popular

Density: 1.5–1.7g/cm³

Hardness: 1–2 on Mohs scale

Oil Absorption Ratio: High; 50–70g of oil per 100g of pigment, resulting in smooth, even texture

Refractive Index: 1.55–1.60

Health and Safety: Non-toxic

Environmental Impact: The plant species luteolin is derived from can have a relatively low environmental impact when sustainably sourced. See also Environmental Impact of Laking Chemicals.

3. In another container, dissolve 10g of sodium carbonate (soda ash) in 250ml of distilled water.

4. Slowly add the sodium carbonate solution to the weld-and-alum mixture while stirring. The sodium carbonate reacts with the alum to precipitate aluminium hydroxide, which binds the yellow dye and forms the weld lake pigment. The solution will become cloudy as the pigment forms. Stop adding the solution when no more visible changes occur.

FILTERING AND WASHING

1. Once the pigment has fully precipitated, filter the solution through a fine muslin cloth or filter paper to collect the weld lake. The pigment will appear as a yellow paste.

2. Rinse the collected pigment with distilled water to remove any remaining soluble salts and impurities, and repeat until the water runs clear. (This is done by flooding the wet pigment with water and stirring whilst in the filter paper/cloth.)

DRYING AND GRINDING

1. Spread the wet pigment on a flat surface or tray and allow it to air dry in a warm, dry place for 2–3 days.

2. Once dry, the pigment can be ground into a fine powder, using a pestle and mortar. Store in a dry place out of direct sunlight.

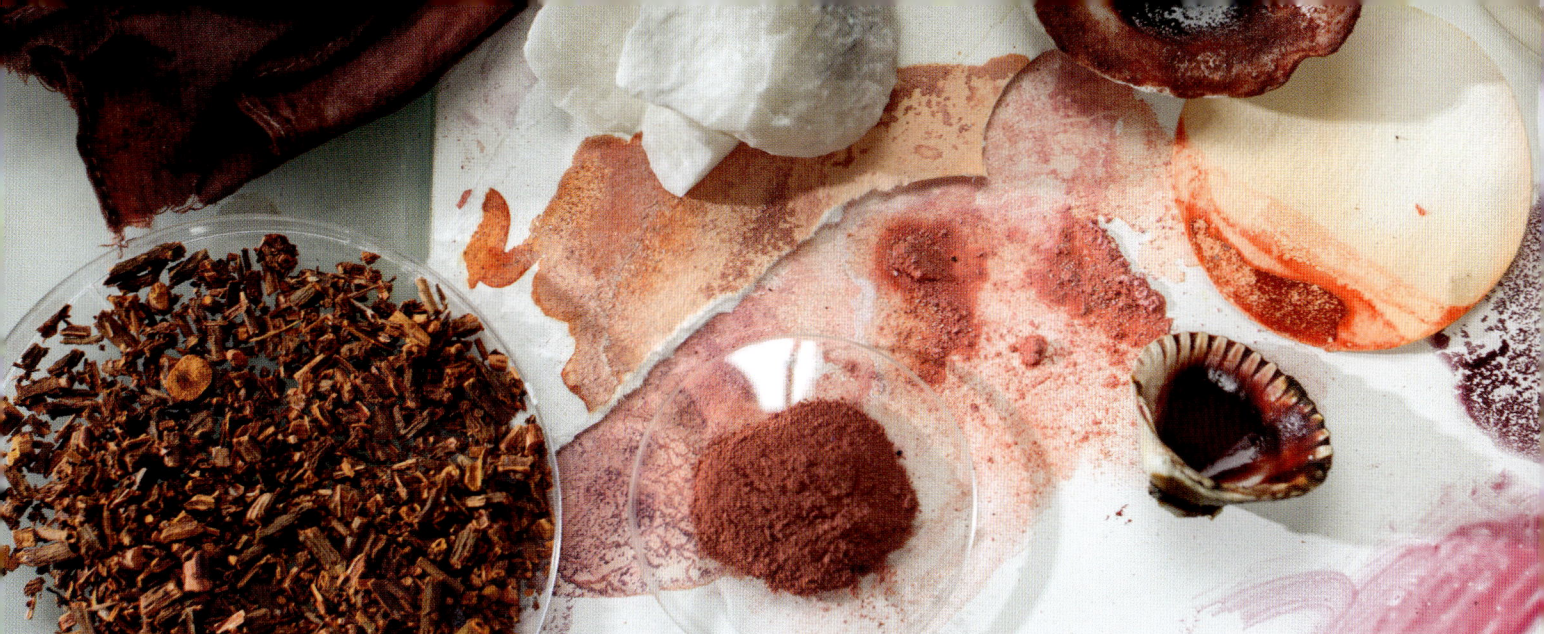

RECIPE: MADDER PIGMENT

Madder is a long-lived perennial of the family Rubiaceae. Madder plants sprout in early April and grow to 60–100cm (2–3¼ft) high. It can be incredibly invasive, so is best grown in pots.

Colourants derived from madder have been used by artists across the globe for thousands of years, as it is one of the most lightfast of the natural dyes. It is known for its rich spectrum of warm tones, ranging from soft pinkish reds to deeper, more intense shades. The plant's cultivation spread across Europe, Asia, and the Middle East, and by the Middle Ages it was a key dyeing agent in the European textile industry. Madder lake pigments became especially popular during the Renaissance when artists sought to create vivid, long-lasting colours for their paintings. The production of a lake pigment from madder roots seems to have been first invented by the ancient Egyptians. Several different techniques and recipes developed over time.

The primary dye component in madder is alizarin, which is responsible for its rich red colour. The best-known source of alizarin is *Rubia tinctorum* (European madder). Used for thousands of years, it has been cultivated in Europe, North Africa, and Asia. Less commonly cultivated is *Rubia peregrina* (wild madder), found across southern Europe and North Africa. Native to the Indian subcontinent and East Asia is *Rubia cordifolia* (Indian madder or Manjistha), widely used for traditional Indian textiles – along with alizarin, it contains purpurin, which imparts orange and red hues. *Rubia akane* (Japanese madder) is found in Japan and surrounding regions and *Rubia argyi* (East Asian madder) in China. In northern Europe, when madder was not readily available, *Galium verum* (lady's bedstraw, yellow bedstraw) was a valuable source of red dye. It has incredibly long, thin roots containing alizarin and purpurin.

Madder dye was considered relatively weak and rather fugitive until 1804, when George Field, the English colour chemist and pigment maker, refined the technique of making a lake from madder by treating it with alum and an alkali. Rose madder supplied half the world with red until 1868, when its alizarin component became the first natural dye to be synthetically duplicated by German chemists Carl Gräbe and Carl Liebermann.

RECIPE 1: MADDER LAKE PIGMENT

Madder lake is made by extracting the red dye (containing alizarin) from madder roots and precipitating it onto a stable base, such as alum, to create an insoluble pigment. The resulting pigment is a bright pink madder lake, known for its transparency and ability to create rich glazes in oil painting and watercolour.

INGREDIENTS

- Dried madder root, 100g
- Alum (potassium aluminium sulfate), 15–20g (20% of the weight of the madder root)
- Sodium carbonate (soda ash), 5–10g (10% of the weight of the madder root)
- Calcium carbonate (chalk), 10g (optional)
- Starch (optional), 5g
- Distilled water
- White vinegar or red wine, 20ml (optional)

EQUIPMENT

- Inert pot (glass or stainless steel)
- Stirring spoon
- Fine sieve or muslin cloth; coffee filter papers
- Glass or plastic containers
- Dehydrator (optional)
- Pestle and mortar

METHOD

EXTRACTING THE DYE

1. Make sure your roots have been washed. Crush or grind dried madder root into coarse lumps, no smaller than approx. 5mm (¼in), as powdered dyestuffs are difficult to filter from the solution. Alternatively, put into a tightly bound muslin bag before soaking.

2. To ferment your plant material, cover it with water to and leave it soaking, at least overnight and up to four days, weighing it down if necessary. See Fermentation for more information. Note: the optional addition of 20ml of vinegar or wine to the initial soak can help to break down the roots and assist with dye extraction.

3. Heat the madder root in distilled water for 2 hours to extract the dye. Impurities from tap water can interfere with the colourants exiting the root and create sadden shades (i.e. muddied colours). Do not exceed 70°C (158°F) as this could denature the dye molecules, so keep the temperature just below. Leave to steep overnight.

4. The next day filter the solution through a fine muslin cloth to remove the root, setting the strained root aside for use in Recipe 2: Rose Madder Extraction. The resulting liquid is our decoction. This should be a deep red wine colour.

PRECIPITATING THE PIGMENT

1. Dissolve alum in warm distilled water in a separate container (typically 10–20% by weight of water; 20g in 200ml of water). (Optionally, 10g of chalk can be added along with the alum to create a more opaque pigment.)

2. Slowly add the alum solution to the madder dye solution while stirring continuously. Continue stirring for a few minutes to ensure the alum thoroughly mixes with the dye. The alum reacts with the alizarin in the madder extract to begin forming the lake pigment.

3. In another container, dissolve sodium carbonate (soda ash) in distilled water to create a basic alkali solution: 10g in 100ml water.

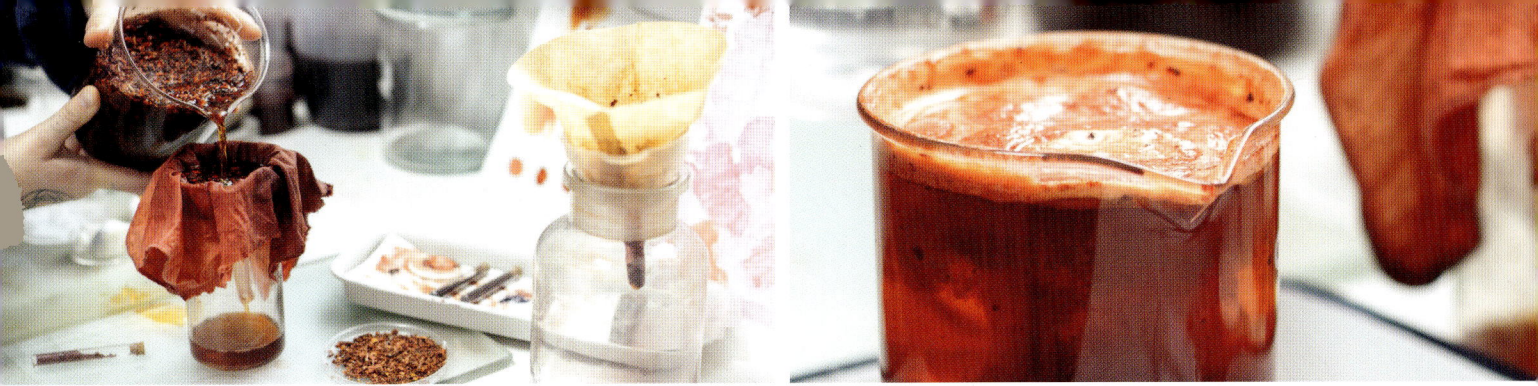

4. Slowly add the alkali solution to the madder-and-alum mixture while stirring. Continue stirring for a few minutes to ensure the sodium carbonate thoroughly mixes with the alum-dye solution. This will cause the aluminium hydroxide to precipitate out and trap the dye, forming the madder lake pigment. The solution will start turning opaque as the lake forms.

5. Continue stirring for 5 minutes. This neutralizes the solution and precipitates aluminium hydroxide, which traps the dye and forms the madder lake pigment. The pigment should start forming as a suspension in the liquid. Effervescence occurs with carbon dioxide gas being given off – this creates a foam with lots of fine bubbles that needs to be 'knocked down' during the reaction.

FILTERING AND WASHING

1. After the pigment has fully precipitated, filter it through fine muslin cloth or a filter paper to collect the solid madder lake.

2. Whilst still in the filter paper or cloth, wash the pigment several times by flooding the pigment with distilled water to remove any remaining soluble salts and impurities. The washing process is crucial for producing a pure, bright pigment that is stable over time. (As it dries, if there are excess reagents or unattached dye molecules you will see this on the filter paper or cloth. The excess dye will dye the cloth and the excess reagents will recrystallize on the surface.)

3. Optionally, 5g of starch can be addded to the washed, wet pigment as a transparent bulking agent.

DRYING AND GRINDING

1. Spread out the filtered madder lake pigment thinly on a flat surface, on the muslin cloth or on the filter paper (split at its seam to lay it flat) and allow it to dry thoroughly. This may take several days; alternatively, use a dehydrator set to 70°C (158°F) for an hour. Try to dry the pigment quickly to stop the growth of mould.

2. Once dry, the pigment can be ground into a fine powder, using a pestle and mortar. Store in a dry place out of direct sunlight.

TIPS

Temperature control: *It is crucial to avoid boiling the madder root during the extraction process; the solution should not reach over 70°C (158°F). Excessive heat can degrade the dye and affect the quality of the pigment, causing it to form a muddier 'red brown' shade, and the resulting pigment will be very hard and brittle.*

Washing the pigment: *Proper washing is essential to remove any remaining alum or sodium carbonate, which could affect the pigment's colour and stability.*

Drying time: *Be patient with the drying process to ensure all moisture is removed before grinding the pigment. Wet or damp pigment can clump and may not mix well with binders.*

pH modifier: *Adding too much vinegar will result in a very hard and larger particle size.*

Quality of roots: *Purchase roots from several suppliers and apply the same recipe to them. Take time to prepare washed and unwashed decoctions to compare colour differences. There is a huge variation in quality from the various suppliers and may be the reason why you do not achieve deep pure red shades.*

RECIPE 2: ROSE MADDER EXTRACTION

This historical recipe illustrates how natural materials like madder root were transformed through various methods to extract a spectrum of vibrant pigments, achieved through careful chemical manipulation. George Field documented a specific technique of pouring an alum solution directly over the madder roots during the extraction process, drawing more of the purpurin content out of the root. This process was intended to maximize the efficiency of the dye extraction and create a richer, more vibrant pigment. His advancements helped improve its stability and vibrancy, enabling artists to achieve much deeper reds compared to earlier preparations. The recipe here is modelled on one that Vanessa Otero from REDiscover demonstrated at the Slade School of Fine Art in 2024 (see Useful Organizations for details and follow the REDiscover project for in-depth research around red lake pigments).

INGREDIENTS

- Set aside madder root from Recipe 1, 10g

- Alum (potassium aluminium sulfate), 5g

- Sodium carbonate (soda ash), 2.5g

- Distilled water

EQUIPMENT

- Linen or cotton fabric

- Funnel, approx 8cm (3in) diameter

- Two 200ml heat-resistant containers

- Inert stirrer (plastic, glass stainless steel)

METHOD

PRECIPITATING THE PIGMENT

1. Cut a piece of linen or cotton into a 25cm (10in) square. Put it inside the funnel so that it overlaps slightly on the circumference, and fold the excess over the rim. Add 10g of the set aside 'spent' madder root from Recipe 1 into the funnel.

2. Dissolve alum in warm distilled water in a separate container (typically 10–20% by weight of water; 5g in 100ml of water).

3. Pour the alum solution slowly over the madder root while agitating with a stirrer. This will help 'catch' as much dye as possible. The alum interacts with the dye compounds (mainly purpurin) in the madder root, helping to fix the colour and form the pigment.

4. In another container, dissolve sodium carbonate (soda ash) in water to create a basic solution. 2.5g in 100ml distilled water.

5. Slowly add the sodium carbonate solution to the madder and alum mixture while stirring the solution constantly. This will cause aluminium hydroxide to precipitate out, trapping the dye and forming the madder lake pigment. The solution will start turning a pink opaque colour as the lake forms.

FILTERING AND WASHING

Refer to Recipe 1: Madder Lake Pigment.

DRYING AND GRINDING

Refer to Recipe 1: Madder Lake Pigment.

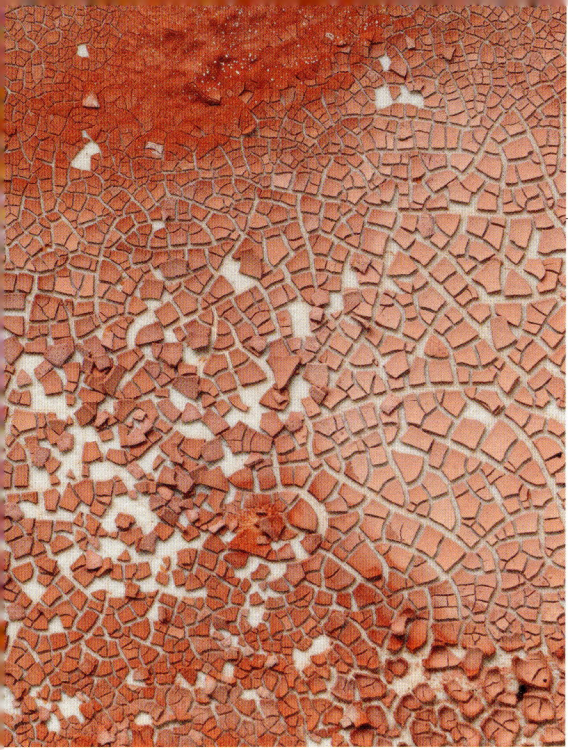

ENVIROMENTAL IMPACT OF LAKING CHEMICALS

Laking chemicals like alum and alkalis (e.g., sodium and potassium carbonate) come from mineral-based raw materials. While their production impacts the environment through mining, energy use, and waste, responsible sourcing and efficient waste management can help reduce harm. By following environmental regulations and sustainable practices, manufacturers can minimize habitat destruction, pollution, and resource depletion. Purchase from reputable retailers; by creating your own lake pigments, you actively support the shift toward more ecologically conscious art practices.

PIGMENT SUMMARY

Pigment Name: Madder lake (made from European madder *Rubia tinctorum*)

Nomenclature: European madder, common madder, madder lake, rose madder, Turkey red, rubia lake, alizarin crimson, madder carmine

Pigment Index Number: NR 9

Colours: Bright orange, red, purple, pink, violet-brown

Colourants & Chemical Formulae: Alizarin $C_{14}H_8O_4$ (red), purpurin $C_{14}H_8O_5$, pseudo purpurin (both orange-red)

Classification: Natural organic

Description: Red lake pigment usually mordanted onto alum

Conservation Issues: Light-sensitive and prone to fading over time if exposed to intense sunlight. Changes occur under different environmental conditions, becoming more orange or brownish with age.

Particle Morphology: Amorphous or fine crystalline structure when precipitated as a lake pigment (usually combined with alum); pigment particles are often fine and irregular

Dates in Use: Ancient times to present. Used for thousands of years, with significant historical importance in the dyeing of textiles and as an artist's pigment in Europe, particularly during the Middle Ages and Renaissance. Its popularity waned after the introduction of synthetic alizarin in 1868.

Density: Typically around 1.5 to 1.8g/cm³, depending on the substrate used for precipitation (e.g., alum)

Hardness: 1–2 on Mohs scale

Oil Absorption Ratio: High; generally requires about 50–80g of oil per 100g of pigment to create a workable paste

Refractive Index: Approx. 1.60 to 1.70, which gives the pigment moderate transparency in thin layers but good opacity in thicker applications

Health and Safety: Non-toxic: madder generally considered safe to handle in normal quantities, though care should be taken to avoid inhaling pigment dust

Environmental Impact: Generally eco-friendly if cultivated responsibly. Excessive use of pesticides or deforestation for cultivation can have negative environmental consequences. Native species could be sought to lower carbon-footprint. See also Environmental Impact of Laking Chemicals.

FERMENTATION

Fermentation of plant tissue plays a crucial role in aiding dye extraction for lake-pigment making, particularly for plants that contain complex dye molecules. It increases the quality and vibrancy of the resulting lake pigment. Whether for madder, weld, or other dye plants, fermentation remains a key step in traditional pigment-making practices. To ferment your plant material, leave it soaking for one to four days in enough water to cover it. If needed, weigh it down to keep it submerged. The following processes will act upon the plant matter to extract as much dye as possible for use in laking.

BREAKING DOWN PLANT CELL WALLS

During fermentation, naturally occurring bacteria, yeasts and fungi break down the tough cellulose structure of the plant cell walls. Invisible to the naked eye, these organisms exist on the surface of plant material and enter the dye liquid through air exposure and contact with equipment. The mechanical breakdown of the plant structure allows the release of dye molecules (such as flavonoids, anthraquinones or indigo precursors) trapped within the plant cells. Fermentation also helps to degrade pectin, a substance that binds plant cells together. This degradation and softening of plant material further facilitates the release of the dye molecules, making extraction more efficient.

CONVERSION OF DYE PRECURSORS

Hidden chemical transformations occur whilst the plant material sits in water. Fermentation aids in the conversion of inactive dye precursors to their active dye forms. For example, in plants like indigo and woad, fermentation is used to convert indicant (a colourless precursor) into indoxyl, which then oxidizes to form the blue pigment indigo.

Below: Fermenting plantstuffs may result in increased brightness of the lake pigment shade.

ACIDIFICATION

As fermentation progresses, organic acids such as lactic and acetic acid are produced. These lower the pH of the environment, which can help transform certain dye compounds, such as flavonoids or anthocyanins, into their more vibrant and stable forms. This process is particularly important for yellow dyes derived from plants like weld or dyer's broom. Fermentation can improve the solubility of certain dye molecules, especially those that are not easily water-soluble in their natural state. The production of organic acids during fermentation alters the chemical environment, making dyes more soluble and easier to extract. This is essential when preparing lake pigments, as the goal is to transfer the dye into a solution.

ENZYME PRODUCTION

Enzymes produced by bacteria and yeasts (such as cellulases and pectinases) that further degrade plant tissues and enhance the release of soluble dyes break down complex carbohydrates (such as starches and pectins) and plant cell walls. This makes it easier to release dye compounds. For instance, in the case of madder root, fermentation helps release alizarin more efficiently from the plant tissues.

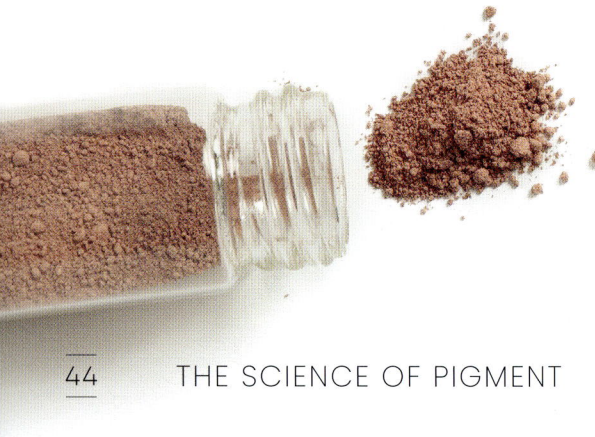

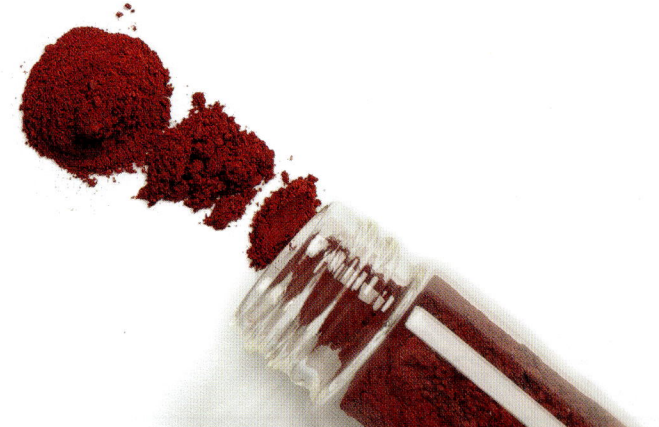

REDUCTION OF UNWANTED COMPOUNDS

The selective breakdown that fermentation helps to reduce leads to the reduction of tannins and other unwanted compounds that may interfere with dye extraction. While tannins can be useful in some dyeing processes, they may darken or dull certain pigments, especially in the production of lake pigments, where the goal is to achieve a pure colour.

ENHANCED PRECIPITATION FOR LAKE PIGMENTS

Fermentation leads to improved dye concentrations in the extract, making it more suitable for lake pigment formation. The breakdown of plant material allows for a higher yield of dye. It also aids in purification of the dye extract by selectively breaking down non-dye components. This results in a cleaner, more vibrant dye solution that is better suited for lake pigment precipitation.

COLOURFASTNESS AND STABILITY

Fermented dye extracts often result in more colourfast and stable lake pigments. This is because the fermentation process can stabilize the dye molecules and eliminate impurities that may lead to fading or degradation over time.

BREAKDOWN OF SUGARS

Sugars in plant material are consumed by microbes during fermentation, which lowers oxygen levels (anaerobic conditions). This anaerobic environment is crucial for the enzymatic breakdown of certain dye precursors like indican in indigo plants, converting them into indoxyl, which can then be oxidized to form the blue pigment indigo upon exposure to air.

The gradual depletion of sugars during fermentation helps to 'clean' the extract, resulting in a purer dye. The breakdown of sugars during fermentation can influence the colour quality of the extracted dye. For example, in plants like madder, fermentation helps convert glycosides (sugar-bound dye precursors like ruberythric acid) into free dyes like alizarin and purpurin.

MICROBIAL ACTIVITY

Lactic acid bacteria (LAB), such as *Lactobacillus* and *Leuconostoc*, are commonly involved in plant fermentation processes. They produce lactic acid by fermenting sugars and carbohydrates found in plant material. This acidification helps break down the plant cells and enhances the release of dye compounds. In addition, the lower pH environment created by lactic acid can activate or stabilize certain dyes, particularly flavonoid dyes like luteolin in weld. Lactic acid bacteria thrive in anaerobic (low oxygen) conditions and at temperatures between 25°C and 35°C (77°F and 95°F).

Acetic acid bacteria (AAB), such as *Acetobacter*, convert ethanol (produced by yeasts) into acetic acid. This acetic acid can help further break down plant material, aiding the extraction of dye. Acetic acid also helps in the conversion of some dye precursors, such as indoxyl in indigo plants, during fermentation. Acetic acid bacteria are aerobic and require oxygen to thrive. They typically operate at temperatures between 20°C and 30°C (68°F and 86°F).

Yeasts such as *Saccharomyces cerevisiae* convert sugars, such as glucose, fructose, and sucrose, present in plant materials into ethanol and carbon dioxide. This metabolic process contributes to breaking down plant material and can alter the chemical environment, facilitating the release of dyes. Yeasts also help create an anaerobic environment that is favourable for the activity of lactic acid bacteria. Yeasts prefer temperatures between 20°C and 30°C (68°F and 86°F) and thrive in the presence of sugars.

Fungi (less common in dye fermentation but can occur), such as *Aspergillus* species, can break down tough plant materials, including cellulose and lignin, helping to release dye molecules. This is more common in composting environments or when fermenting harder plant materials. Fungi often require aerobic conditions and operate best at slightly lower temperatures (15°C to 25°C/59°F to 77°F).

SPEEDING UP THE FERMENTATION PROCESS

Fermentation is naturally slow, but the process can be accelerated by optimizing several factors:

Increasing Temperature: Warmer temperatures speed up microbial activity. Fermentation typically progresses faster between 25°C and 35°C (77°F and 95°F). However, care must be taken not to exceed optimal temperature ranges, as too much heat can denature enzymes and halt the fermentation process.

Introducing Starter Cultures: Using a starter culture of lactic acid bacteria (such as from sauerkraut juice or yogurt whey) can accelerate the fermentation process. Starter cultures contain high concentrations of active bacteria that can quickly initiate fermentation. For instance, adding a *Saccharomyces cerevisiae* yeast starter can speed up ethanol production, which then feeds acetic acid bacteria in the case of indigo fermentation.

Controlling pH: Adding small amounts of an acid, such as white vinegar, can pre-acidify the solution and encourage faster growth of lactic acid bacteria while inhibiting unwanted microbial growth. In some cases, a slightly acidic environment can also help to release dye compounds more efficiently.

Controlling Oxygen: For anaerobic bacteria like lactic acid bacteria, ensuring that the plant material is fully submerged in the fermentation liquid helps create the necessary low-oxygen environment. Conversely, for aerobic processes (such as those involving acetic acid bacteria), ensuring adequate exposure to oxygen will speed up fermentation.

Using Smaller Plant Pieces: Chopping or grinding the plant material into smaller pieces increases the surface area available for microbial activity and dye release. This can significantly reduce the time needed for fermentation.

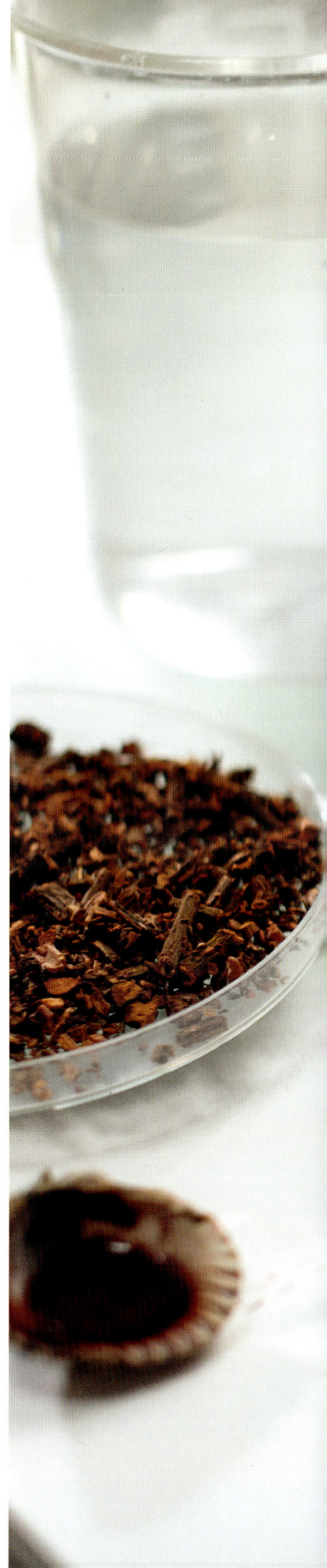

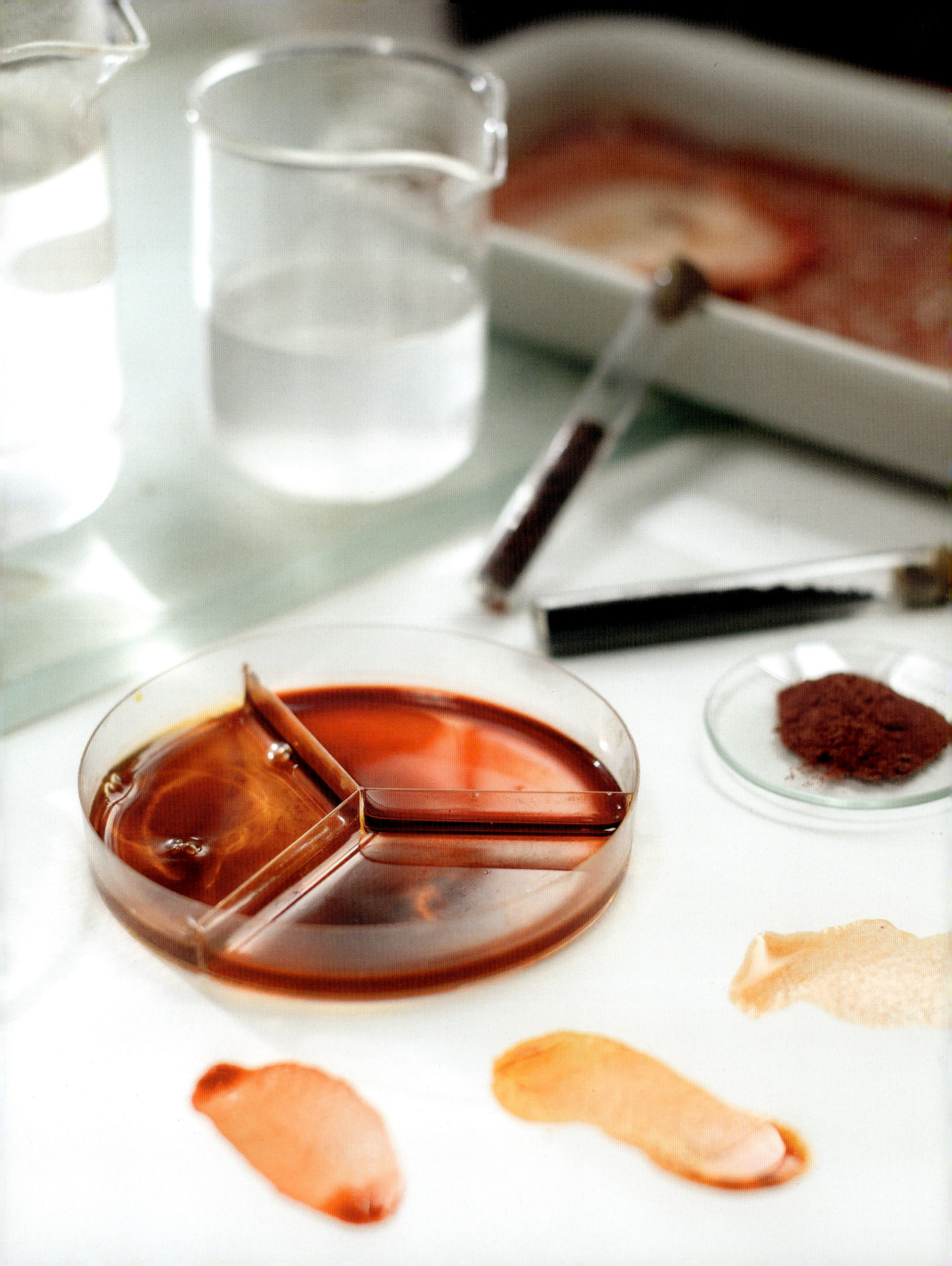

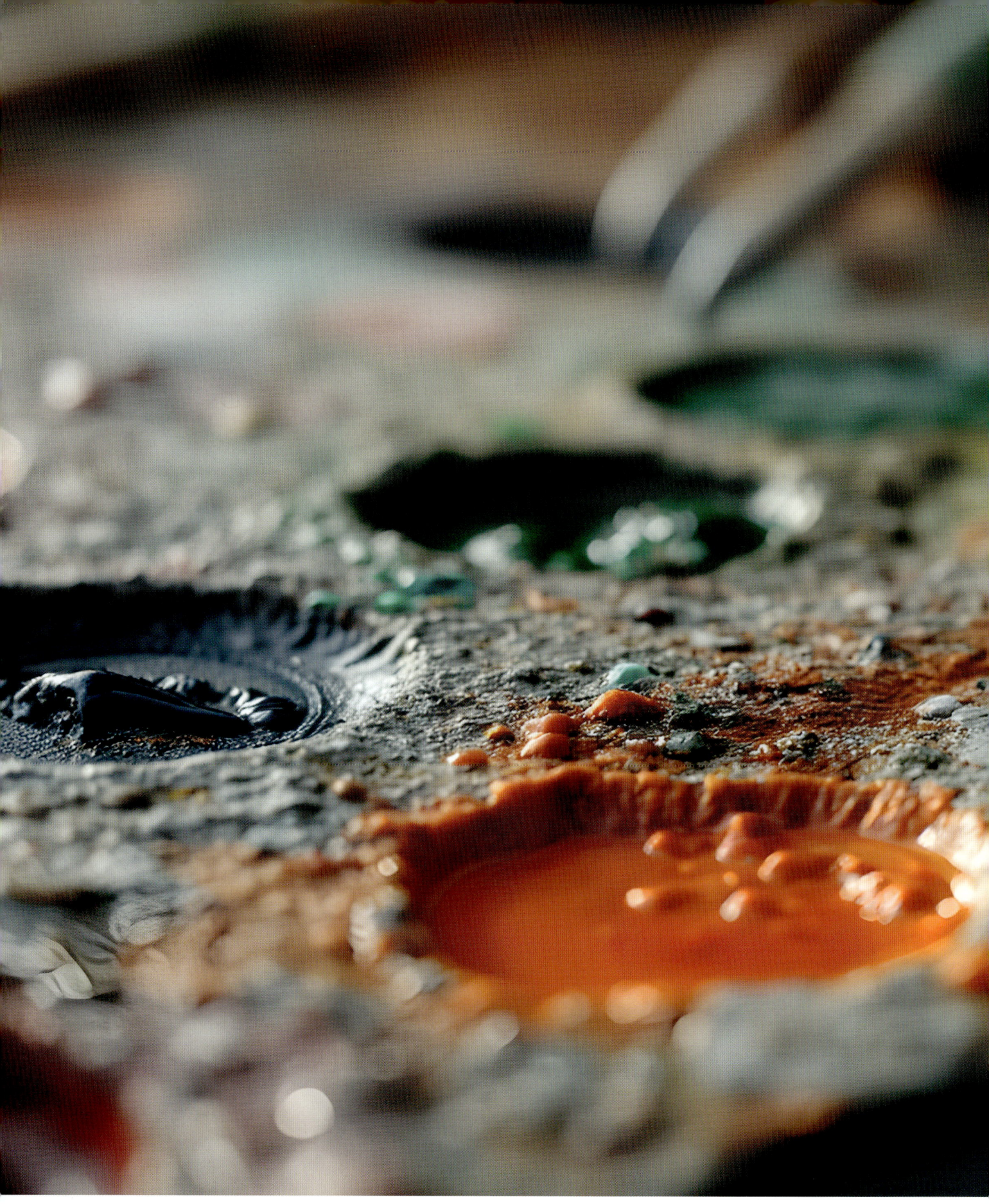

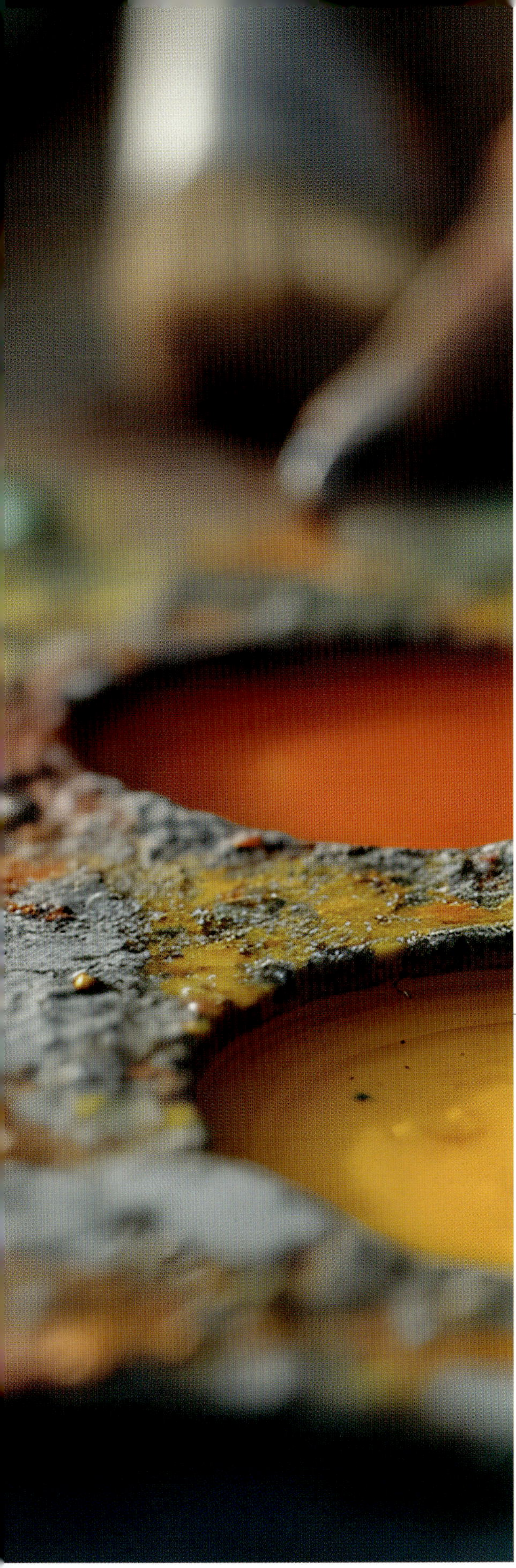

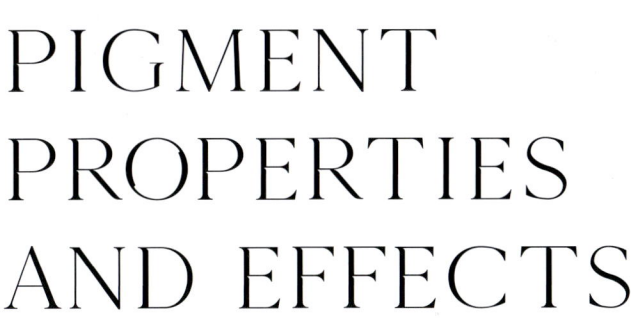

CHAPTER THREE

PIGMENT PROPERTIES AND EFFECTS

Emphasizing its importance as a colourant, the word 'pigment' comes from the Latin *pigmentum*, derived from *pingere*, to paint. From tiny humble particles, important ideas and emotions are conveyed. Pigments become more than what nature made; in the hands of the artist they become a means of communication. With insight into pigment behaviour, creatives can optimize pigment effects, refining their technique to ensure the material physically embodies their intent – honing their craft through careful manipulation. When the material physically embodies the artist's purpose, the artwork transcends mere technique, becoming an extension of thought, emotion, and vision.

UNDERSTANDING PIGMENT PROPERTIES

As seen in Chapter 2, chemical reactions within pigments can alter their colour or composition. For example, pigments that contain sulfur, such as cadmium yellow, can react with airborne pollutants such as hydrogen sulfide to form black cadmium sulfide, thus darkening the colour.

Certain pigments are vulnerable to acidic environments, causing them to fade or change colour. These changes are characterized as 'chemical stability', be it stronger or weaker. A pigment's ability to withstand high temperatures without decomposing or altering its colour is referred to as its 'thermal stability'. This is particularly important in industrial applications such as ceramics and plastics, where pigments may be exposed to high heat during processing.

A pigment's resistance to fading when exposed to light, particularly ultraviolet (UV) radiation, is measured as 'lightfastness'. Organic pigments are often more susceptible because the energy from light can break down their molecular structure, leading to fading. In contrast, many inorganic pigments, especially those based on metal oxides, exhibit excellent lightfastness due to their stable, crystalline structures.

Pigments are mixed with binders and mediums to create paints, inks, and other materials. These can significantly impact the behaviour of the pigment, including its colour intensity, texture, drying time, and overall stability. A binder holds the pigment particles together and adheres them to the surface. Common binders include oils (in oil paints), gum arabic (in watercolours), and acrylic polymers (in acrylic paints). The interaction between pigment and binder can affect the final appearance of the colour – for example, oil tends to enhance the depth and richness of pigments, while water-based binders may produce more transparent and delicate hues.

Mediums can be added to the pigment-binder mixture to alter the properties of the paint. These might include solvents, thickeners or drying agents. The medium can influence the pigment's dispersion, flow, and drying time, which in turn affects how the pigment behaves when applied to a surface.

The shape, size, and surface characteristics of pigment particles are collectively termed 'particle morphology'. These attributes create a unique set of behavioural characteristics that play a crucial role in determining the physical and optical properties of pigments. Advances in pigment technology continue to explore new morphologies to achieve desired effects in increasingly innovative and sustainable ways. Whether in fine art, industrial applications or cosmetics, the morphology of pigment particles plays an essential role in creating the colours that define our world.

Right: Pigment in dry form, as a wetted watercolour and as a dry colour swatch.

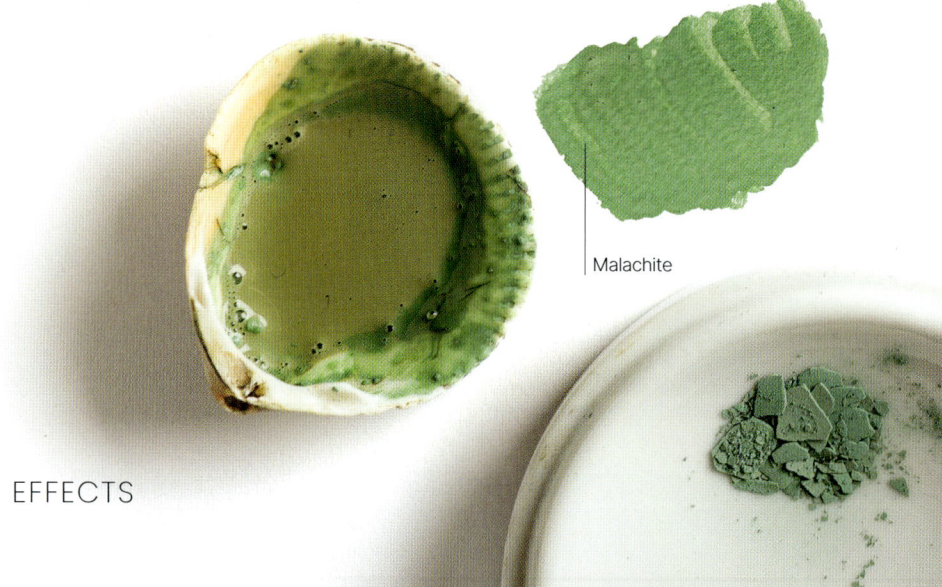

Malachite

Understanding the different particle shapes, from spherical to fibrous, allows manufacturers and users to select the appropriate pigment for their specific needs, optimizing properties such as colour strength, opacity, durability, and texture.

The optical behaviour of individual pigments is also affected by particle size. Early and historical pigments are constituted from particles that vary in size, as they would have been ground by hand, whereas modern pigment particles are more uniform due to being mechanically ground or chemically formed. Particles that are uniform have higher coverage – this is typical of modern synthetic pigments, and quite unlike the unique qualities of the more esoteric and subtle early pigments.

The way in which a pigment is structured determines the finish of the paint in terms of an even consistency, homogenous smoothness and gloss of the paint layers. A smaller homogenous grain size will produce smoother paint surfaces, with more homogenous colour, as it reflects light evenly. If burnished, it can become shiny.

Blues and greens such as azurite and malachite derived from naturally occurring inorganic sources are especially reliant on particle size to determine the intensity of the colour. Particles of a large size impart very saturated colour but have less tinting strength than finely ground pigment. The importance of the ground colour will therefore have greater effect with coarsely ground pigment.

Pigment manufacturers and paint makers add fillers and other additives to create colour ranges that are homogenous in nature. The individual character of pigments can be lost in the processing. By making them by hand, one can become familiar with the behaviour of different source materials. Intimate connections and moments with colour can be formed when you collect it and make it yourself.

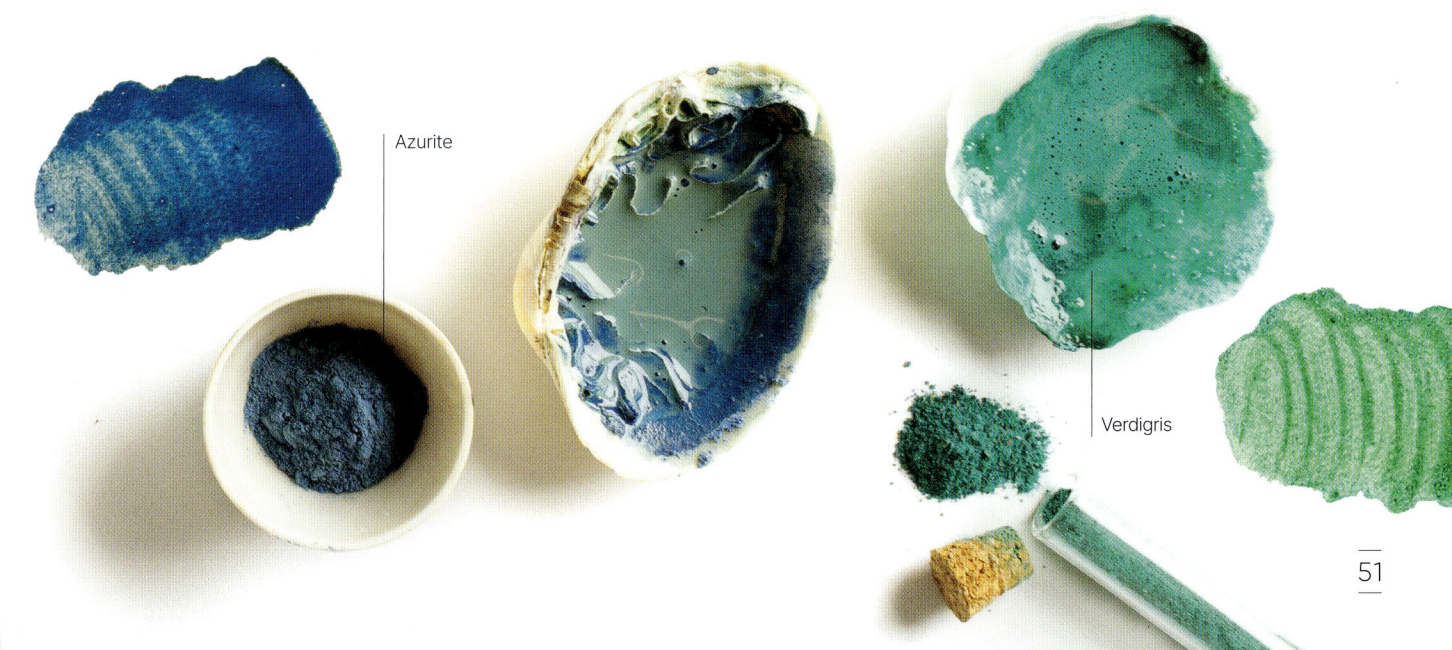

Azurite

Verdigris

PARTICLE SHAPES COMMONLY FOUND IN PIGMENTS

SPHERICAL PARTICLES

Characteristic round, ball-like shape. They tend to have a uniform distribution in size, which leads to consistent optical properties, enhancing the opacity of the pigment. Spherical pigments generally have good flowability, making them easy to disperse in a medium. Due to their shape, they can provide a smooth surface finish, contributing to higher gloss in paints and coatings.

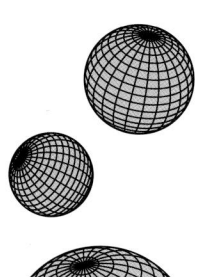

EXAMPLES

Titanium Dioxide (TiO$_2$): Often used in its spherical form for maximum opacity and brightness in paints.

Synthetic Iron Oxides: Can be produced with spherical morphology to optimize dispersibility and colour uniformity.

PLATELET OR FLAKE-SHAPED PARTICLES

Flat and thin, with a high aspect ratio. They resemble small plates or flakes, and are often transparent or semi-transparent. Excellent at reflecting light, they are ideal for metallic and pearlescent effects in coatings and cosmetics. Due to their shape, they can create overlapping layers that enhance the barrier properties of coatings, improving resistance to moisture and chemicals. Interesting orientation effects can be achieved; platelet pigments can align parallel to the substrate in coatings, enhancing the smoothness and gloss of the surface.

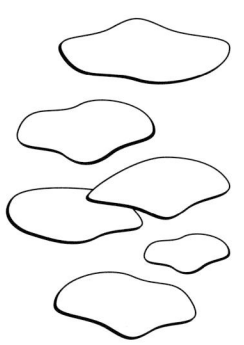

EXAMPLES

Mica-Based Pearlescent Pigments: Used in automotive and cosmetic applications for their shimmer and lustre.

Aluminium Flakes: Often used in metallic paints and coatings for a shiny, reflective finish.

Vermilion (Cinnabar): Used in historic artworks in its natural platelet form for vivid red hues.

Clays: Alumina silicates such as kaolite and chlorite.

ROD-SHAPED PARTICLES

Elongated and cylindrical, resembling tiny rods or needles. The aspect ratio (length to width) of these is typically greater than 1. Rod-shaped particles exhibit anisotropic properties (i.e., having different physical properties in different directions) that can vary depending on the direction of measurement. This can lead to unique optical effects such as anisotropic scattering. In applications such as coatings, rod-shaped particles can align in a particular direction, influencing the texture and mechanical properties of the coating. They may require specific dispersing techniques to prevent them from aligning or clumping, which could affect the final appearance.

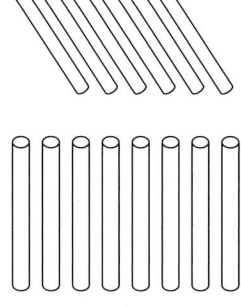

EXAMPLES

Carbon Black: Sometimes found in a rod-like form; used in inks and coatings for deep black colours.

Some Organic Pigments: Certain azo pigments, for example, can have a rod-shaped morphology, affecting their gloss and dispersion characteristics.

ACICULAR (NEEDLE-LIKE) PARTICLES

Similar to rod-shaped particles, but typically thinner and more elongated, often resembling needles. In certain applications, such as composite materials, acicular pigments can provide mechanical reinforcement due to their shape. The needle-like structure allows particles to interlock, providing enhanced structural integrity in coatings or plastics; however, acicular pigments can be challenging to disperse evenly without specialized techniques.

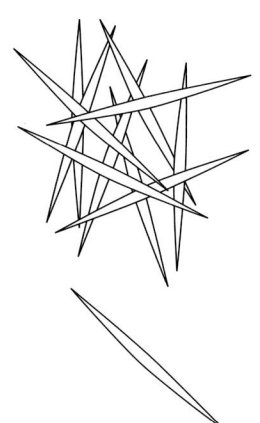

EXAMPLES

Rutile Titanium Dioxide: Can form acicular particles that provide superior opacity and UV protection in coatings.

Some Hematite-Based Pigments: Used in industrial applications where both colour and mechanical strength are required.

IRREGULAR OR AGGREGATED PARTICLES

Lacking a defined shape, these often appear as jagged or aggregated clusters of smaller particles. Aggregation occurs when primary particles clump together during production or storage. Irregular particles typically have a high surface area, which can enhance their colour strength and reactivity. Their irregularity can scatter light diffusely, resulting in a matte finish in coatings or paints. Aggregated particles can create porous structures, influencing absorption properties and how the pigment interacts with binders or other chemicals.

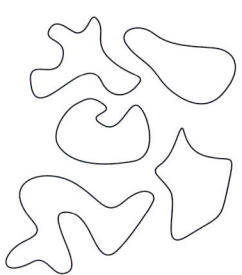

EXAMPLES

Natural Earth Pigments: Due to natural grinding and processing methods, many have irregular shapes, such as ochres and umbers.

Lake Pigments: Artisanally crafted lake pigments, unlike those produced in a lab, form irregularly shaped amorphous particles.

HOLLOW SPHERES

Comprise an outer shell with an empty core. These particles are engineered to have specific optical and physical properties. The hollow structure makes them lightweight, which can be beneficial in applications like aerospace coatings; they scatter light effectively, providing good opacity with less material; and air trapped inside the hollow spheres can provide thermal insulation, useful in certain industrial coatings.

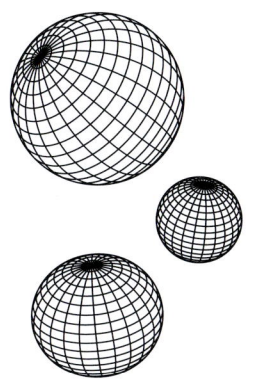

EXAMPLES

Glass Microsphere: Used in reflective paints and coatings, especially for road markings and safety signs.

Ceramic Hollow Spheres: Applied in high-temperature coatings where both insulation and lightweight properties are needed.

FIBROUS PARTICLES

Consist of long, thin strands that resemble fibres. Not common in traditional pigments, but used in specific applications where texture or reinforcement is needed. Fibrous pigments add mechanical strength to materials like plastics and coatings, and they can create a textured surface when used in paints or coatings, providing unique aesthetic effects. However, the paint finish can be affected by the particle orientation during application, significantly affecting the final appearance and mechanical properties of the material.

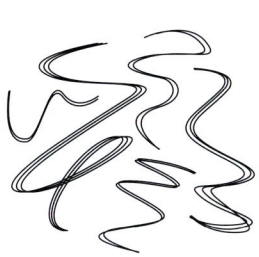

EXAMPLES

Asbestos Pigments: Historically used in various industrial applications; now largely discontinued due to health risks.

Fibrous Clays: Certain clays with fibrous morphologies are used in ceramics and speciality coatings.

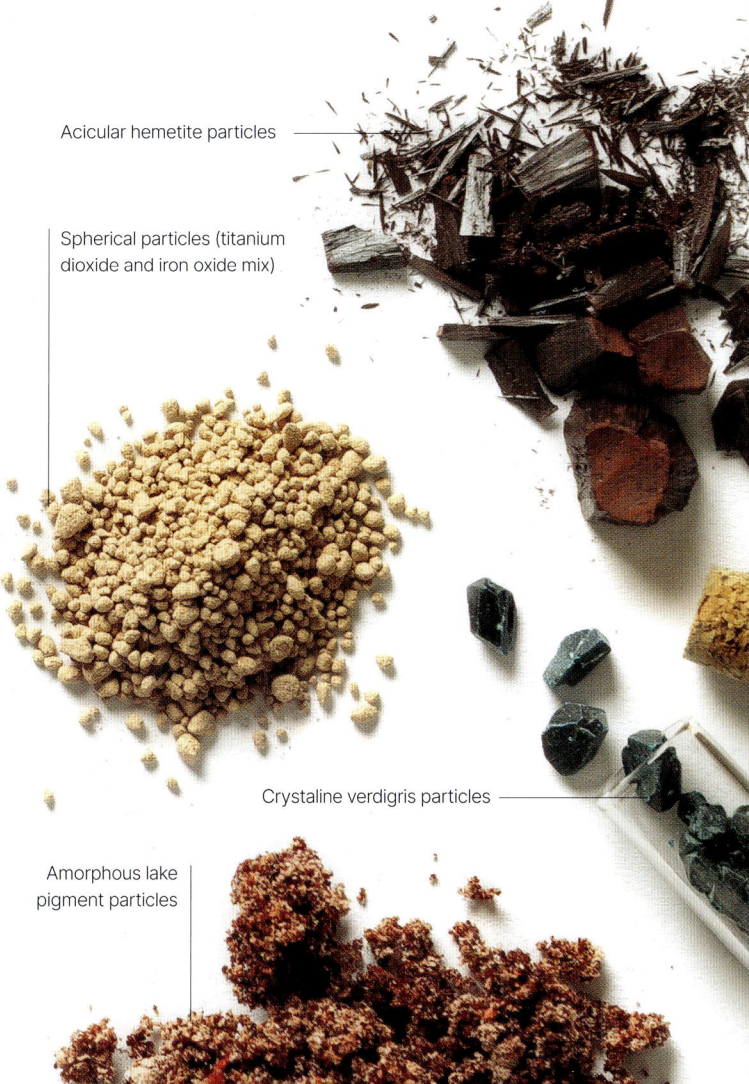

Acicular hematite particles

Spherical particles (titanium dioxide and iron oxide mix)

Crystaline verdigris particles

Amorphous lake pigment particles

Ox gall

Gum Arabic crystals (Nigeria)

INCREASING GRANULATION IN YOUR PAINT

Granulation is the term used for pigment particles clumping together into larger, visible clusters within the paint. Knowing the science behind this can inform artistic choices. Here are some additives and techniques you can use to promote the clustering of pigment particles and thereby increase the granulation in your paint. This is especially useful in water-based mediums such as watercolour – where, for example, granulation can create a textured, speckled effect on the paper. Smoky, brooding, and atmospheric washes can also be achieved.

GUM ARABIC

By slightly decreasing the concentration of this natural binder in your paint mixture, you can enhance the pigment's tendency to settle into the paper's texture, thereby promoting granulation. This is because gum arabic affects how pigment particles move and settle as the water evaporates.

ISOPROPYL ALCOHOL OR RUBBING ALCOHOL

When alcohol is applied to wet paint, it repels water, pushing pigment particles away and causing them to cluster in some areas, creating a granulation effect. This method is often used to create dramatic textures in watercolour.

LAYERED WASHES

By applying multiple thin washes of paint, allowing each layer to dry before adding the next, the granulating pigments in successive layers will accumulate in the paper's texture, enhancing the granulation effect.

OX GALL

Ox gall, a liquid extracted from the gall bladder of cattle, is traditionally used in watercolour painting to reduce the surface tension of the water in the paint. This allows pigment particles to separate and settle more easily. The resulting uneven distribution of pigments leads to a more granulated effect.

TABLE SALT OR COARSE SEA SALT

When sprinkled onto wet paint, salt absorbs water and pigment, pulling the pigment particles into small, concentrated areas. This creates localized granulation and can produce a textured, crystalline effect. It's a common technique in watercolour painting to create varied textures.

PIGMENT CHOICES

Use pigment mixtures that contain particles of a great diversity of size. Very small pigment particles will be attracted by larger ones, which upon drying will produce incredible organic patterns in the paint. Using irregularly shaped pigment particles will emphasize this affect.

Coarse sea salt

Granulation caused by impurities in tap water

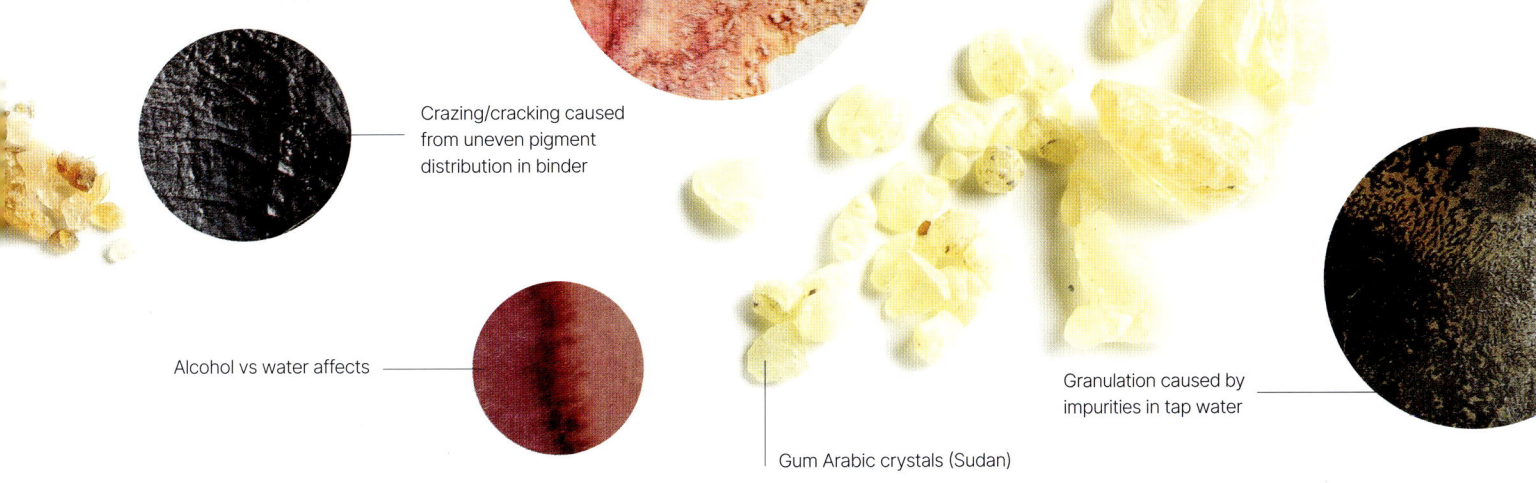

Crazing/cracking caused from uneven pigment distribution in binder

Alcohol vs water affects

Granulation caused by impurities in tap water

Gum Arabic crystals (Sudan)

VARIED WATER QUALITY

By mixing some pigments with distilled water and some with tap water, interesting granulating affects can occur when the washes or paints meet.

WATER WITH HIGH MINERAL CONTENT (HARD WATER)

Hard water accelerates granulation of pigment particles due to its high content of dissolved minerals, primarily calcium (Ca^{2+}) and magnesium (Mg^{2+}) ions. As the surface charges are neutralized by the calcium and magnesium ions, the repulsive forces between the pigment particles weaken. This allows the particles to come closer together and clump more easily. In addition, calcium and magnesium ions in hard water act similarly to flocculants (see The Science of Particle Clustering in Chapter 2). Tap water contains impurities such as chlorine and hard water contains calcium compounds, both of which aid granulation. Hard water can also slightly alter the surface tension of the water, affecting how pigments interact with the medium. The resulting clumps settle unevenly on the textured surface of the paper – darker areas of pigment collect in small clusters, creating a mottled effect.

CONSERVING GRANULATING EFFECTS

While granulating paints are not necessarily more chemically unstable, their textured nature and uneven distribution can make them more vulnerable. Conservation measures can mitigate these risks and ensure the longevity of granulating paint-based artworks. These include careful handling during cleaning or treatment, as loosely bound pigment particles can be more easily dislodged.

Where granulating paint shows signs of flaking or powdering, conservators may need to use consolidants to stabilize the pigment. These are carefully chosen to preserve the distinctive texture of granulating paints. Artworks using granulating paints are often framed under glass to protect them from dust, moisture, and mechanical damage. Maintaining stable environmental conditions (controlled humidity and temperature) is crucial for the long-term preservation of these works (see Factors Affecting Stability of Granulating Paints).

Layered washes to increase granulation

Darker side of swatch caused by iron impurities in tap water

Irregular pigment particles can cause granulation affects

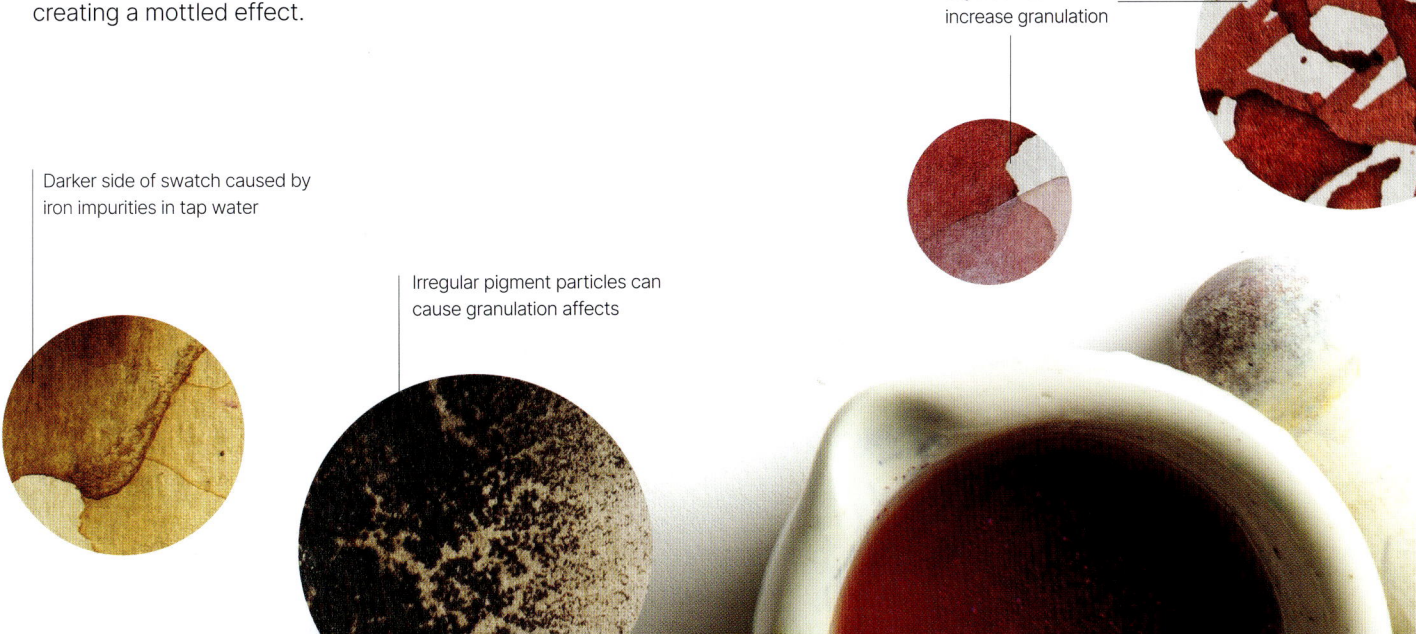

FACTORS AFFECTING STABILITY OF GRANULATING PAINTS

PIGMENT COMPOSITION

Natural Earth Pigments: Many granulating pigments are naturally occurring earth pigments (e.g., ochres, umbers), which are generally stable over time due to their inorganic nature. These pigments have been used for centuries and tend to be chemically inert, making them relatively durable.

Heavy Mineral Pigments: Some granulating pigments, like cobalt or manganese-based pigments, are more prone to degradation under certain environmental conditions, such as exposure to light, moisture, or acidic environments. However, these are typically stable under normal conservation conditions.

PARTICLE DISTRIBUTION

Granulating pigments create uneven surfaces due to their clustering and settlement. This can lead to areas where the pigment is less tightly bound to the substrate (e.g., paper or canvas). Such areas may be more prone to mechanical damage (abrasion or flaking) because the pigments are sitting loosely on the surface rather than being uniformly embedded in the medium.

CHOICE OF BINDING MEDIUM

Watercolour And Gouache: Granulating pigments are often used in watercolours and gouache, where the binder (typically gum arabic) is relatively weak compared to oil or acrylic paints. This makes the paint layer more sensitive to water, humidity, and physical contact and microbial attack, which could lead to pigment loss over time.

Oil And Acrylic: In oil or acrylic media, the binding strength is usually higher, providing more protection to the granulated pigment. However, poor formulation or degradation of the binder (e.g., cracking or yellowing of oil) could still expose the granulated pigment to damage.

LOW BINDING IN CERTAIN AREAS

Since granulating pigments settle in clusters, some areas of the paint layer may have thinner applications of the binding medium. These areas are potentially more vulnerable to environmental degradation, such as dust accumulation, and are more likely to have less binder to pigment ratio resulting in flaking. Thinner areas of pigment application are liable to be affected by UV light and fade more quickly.

ENVIRONMENTAL CONDITIONS

Humidity And Moisture: Granulating pigments, especially in watercolour, can be more vulnerable to moisture, as water can redissolve the binder (gum arabic) and lead to pigment movement or loss. High humidity environments may increase the risk of mould or other biological degradation, particularly on textured surfaces where granulating pigments have settled unevenly.

Light Sensitivity: Granulating pigments, particularly those based on heavy metals (e.g., cobalt, manganese), can be sensitive to light over long periods. Prolonged exposure to UV light may cause some pigments to fade, go darker or change colour, especially in watercolours, where the pigment is more exposed.

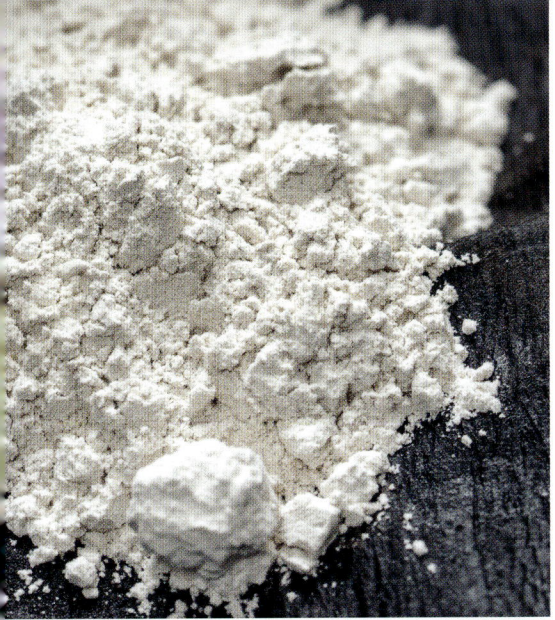

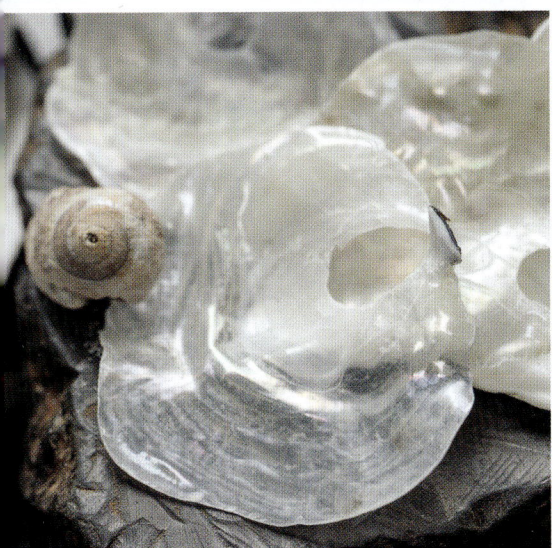

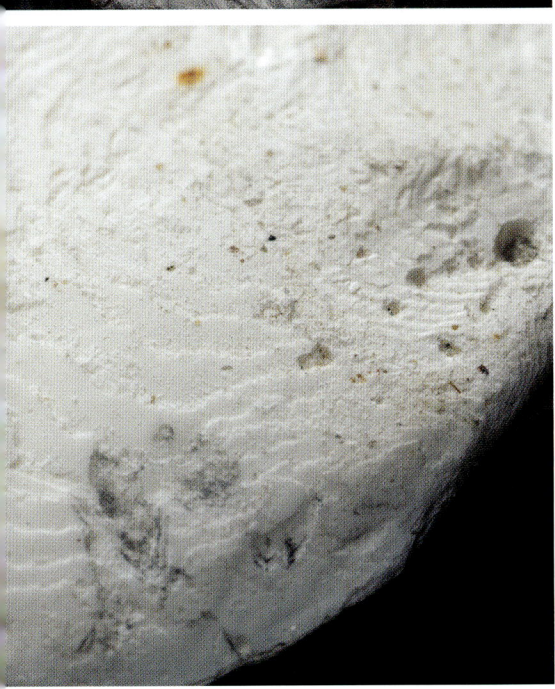

AN ARTIST'S ASIDE

EXTENDER MATERIALS AND THOUGHTS ON CHALK

White extender pigments are added to paints to lower their cost or improve their properties. They include calcium carbonate, calcium sulfate, diatomaceous silica (the remains of marine organisms), china clays, and fumed silica. Calcium carbonate contributors – including plankton (such as coccoliths and planktic foraminifera), coralline algae, sponges, brachiopods, echinoderms, bryozoa, and molluscs – are typically found in shallow-water environments where sunlight and filterable food are more abundant. Eggshells, snail shells, cuttlefish, and seashells are sources for white pigments that contain calcium.

The Old Saxon word for chalk is *cealc*. This word slivers, a soft beginning, ending in a crunch – that compact crackle that deep snow has. *Cealc* is a greasy lubricant, a velvety mattifier, and the paint maker's secret weapon, to enliven and reveal hidden mid-tones. Chalk on the other hand sounds a little harsher, a small percussion. The rhythms emulate its birth, a repeated building up of sediment, over and over again, compacted and condensed and compressed into dense white matter. The individual resonance of these two words emphasizes the different properties of these materials through the conjuring of different sounds and images.

Rounded pebbles made of chalk that can be found on the Thames foreshore are known as 'Thames spuds'. Their softly curved bodies reveal their forced undulating movement, victims of the quick erosion of the London tide. These tumbled pieces are reminiscent of the potatoes they are colloquially named after – they even have the same weight. When wet, chalk can sometimes have a gloopy starchiness too.

The earliest lake pigments, dyes derived from plant or animal origin, have for hundreds of years been chemically attached to chalk. It acts as both a substrate particle and 'chemical fix' for the colour and gives the fugitive pigments a slightly longer life.

A collaboration with nature. Just imagine – million-year-old plankton and molluscs, whose shells are now used to keep and hold onto colour, keeping it safe for us to appreciate these fleeting colours just a little longer.

CREATING IRON OXIDE SYNTHETICALLY

Synthetic iron oxide can be produced both industrially and at home through chemical processes that result in the creation of iron oxide pigments. These pigments come in various colours (red, yellow, black, brown) and are used in industries such as paints, coatings, plastics, and construction due to their durability, opacity, and resistance to weathering.

Industrial processes such as the precipitation method and thermal decomposition create pigments on a large scale, while at-home recipes can produce smaller quantities of pigments through simple reactions using common materials. An artist might want to understand the different ways of making iron oxide pigments, both synthetically and from naturally occurring sources, for several important reasons that enhance their creative practice, technical knowledge, and appreciation of materials. The chemical reactions underlying these processes involve the transformation of iron in the presence of oxygen and various acids or alkalis, resulting in vibrant, durable pigments that have been used for centuries in art and industry.

METHODS FOR INDUSTRIAL PRODUCTION OF SYNTHETIC IRON OXIDE

There are several methods for producing synthetic iron oxide on an industrial scale. The two most common are given here.

METHOD 1: PRECIPITATION (PENNIMAN-ZOPH PROCESS)

This is the most common method for producing synthetic iron oxides, particularly for red, yellow, and black iron oxides. The process involves the controlled oxidation of iron in an aqueous medium.

TO MAKE YELLOW IRON OXIDE (GOETHITE, $FeO(OH)$)

This is produced by the controlled hydrolysis of ferrous sulfate in the presence of an alkali (usually sodium hydroxide).

First, ferrous sulfate ($FeSO_4$) is dissolved in water. Then sodium hydroxide ($NaOH$) is added to precipitate iron hydroxide. The solution is heated. Oxygen is introduced, converting the iron hydroxide to yellow iron oxide (goethite).

TO MAKE RED IRON OXIDE (HEMATITE, Fe_2O_3)

This is produced by oxidizing yellow iron oxide at higher temperatures (approximately 600°C/1112°F). This converts the goethite ($FeO(OH)$) into hematite (Fe_2O_3), which is red in colour.

TO MAKE BLACK IRON OXIDE (MAGNETITE, Fe_3O_4)

This is produced by partial oxidation of ferrous salts under controlled conditions.

Ferrous sulfate is dissolved in water, and the solution is treated with an alkali to precipitate ferrous hydroxide ($Fe(OH)_2$). Controlled oxidation (under low oxygen conditions) converts the ferrous hydroxide to magnetite (Fe_3O_4), which is black and also magnetic.

METHOD 2: THERMAL DECOMPOSITION

This method is used primarily for the production of red iron oxide (hematite).

Iron sulfate or iron chloride is heated to temperatures around 600–700°C (1112–1292°F). These compounds decompose, releasing sulfur dioxide or chlorine gas, and leaving behind iron oxide.

RECIPE: MAKING SYNTHETIC IRON OXIDE AT HOME

While producing synthetic iron oxide at home is less efficient than industrial methods, it can be done using simple materials like steel wool, vinegar, and hydrogen peroxide. The process typically results in red iron oxide (rust, Fe_2O_3), but with modifications, black and yellow oxides can also be produced.

Leaving metal to rust in water, acetic acid, and household vinegar in controlled conditions catalyzes the rust that would naturally occur in non-museum conditions, to create a range of yellow, red, and brown pigments. Some of these solutions are used in natural dyeing processes as metal mordants to help 'fix' natural dyes to the textile.

RECIPE 1: RED IRON OXIDE (RUST)

INGREDIENTS & EQUIPMENT

- Steel wool
- White vinegar (acetic acid)
- Hydrogen peroxide
- Plastic or glass container

METHOD

1. Soak the steel wool in white vinegar for a few hours to dissolve the protective coating on the steel and initiate the oxidation process.

2. After the steel wool has soaked, remove it from the vinegar and place it in a container with hydrogen peroxide. The hydrogen peroxide acts as a strong oxidizer and accelerates the rusting process.

3. Over time, the steel wool will rust, forming red iron oxide (Fe_2O_3). This process can take a few days to a week.

4. Once the steel wool has fully rusted, let it dry. After drying, grind the rust into a fine powder for use as a pigment.

RECIPE 2: BLACK IRON OXIDE (MAGNETITE)

INGREDIENTS & EQUIPMENT

- Steel wool or iron filings
- White vinegar (acetic acid)
- A low-oxygen heat source (e.g., a closed oven)
- Plastic or glass container

METHOD

1. Similar to the method for red iron oxide, steel wool is soaked in vinegar to begin the oxidation process.

2. After soaking, the steel wool is heated in an oven at a low temperature with limited oxygen access (wrapping the steel wool in aluminium foil, for example) to encourage the formation of black magnetite instead of red rust. (Alternatively, try this in a kiln in a reduction atmosphere.)

3. The steel wool should turn black as magnetite (Fe_3O_4) forms.

RECIPE 3: YELLOW IRON OXIDE (GOETHITE)

INGREDIENTS & EQUIPMENT

- Iron salts (e.g., iron sulfate or iron chloride), 7g
- Baking soda (sodium bicarbonate) or chalk, 10g
- Distilled water
- Plastic or glass container

METHOD

1. Mix iron salts (7g) and baking soda (or chalk) (10g) with 50ml distilled water each in separate solutions.

2. Combine solutions and stir well. This will cause iron hydroxide ($Fe(OH)_3$) to precipitate out. Alternatively use a 1:1 ratio of iron salts to alkali to achieve a more opaque pigment.

3. Allow the iron hydroxide to oxidize slowly in the air. This process will form yellow iron oxide (goethite, $FeO(OH)$).

4. Once fully oxidized, dry the yellow pigment and grind it into a fine powder.

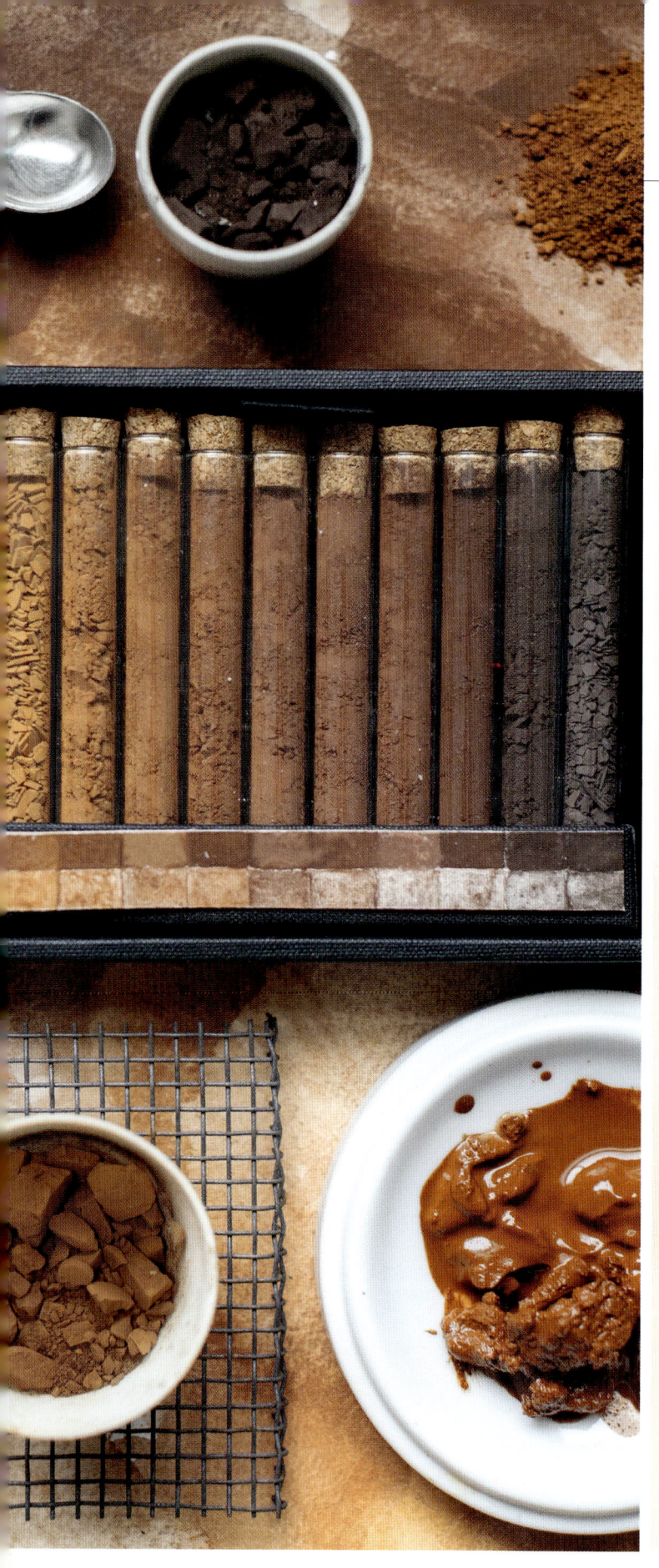

PIGMENT SUMMARY

Pigment Name: Synthetic iron oxides: yellow (Y), red (R), black (B), naturally occurring as goethite, hematite, magnetite respectively

Nomenclature: (Y), (R), (B) ochres, Mars colours, ferric oxide pigments; (R) Indian red, Venetian red; (B) mineral black

Pigment Index Number: (Y) PY 42; (R) PR 101; (B) PbK 11

Colours: Yellow, red, black (orange and brown if mixed)

Chemical Name: (Y) hydrous iron oxide; (R), (B) anhydrous iron oxide

Chemical Formula: (Y) $FeO(OH)$; (R) Fe_2O_3; (B) Fe_3O_4

Classification: Synthetic inorganic

Description: Highly stable, non-toxic, durable; produced by a controlled process of heating iron compounds (e.g., iron salts) in the presence of oxygen, resulting in various particle shapes (spherical, cubic, acicular) – characteristics that contribute to strong tinting power and high opacity

Conservation Issues: Minimal degradation; not prone to fade or alter chemically in most environments; safe for archival use

Particle Morphology: (Y) acicular or irregular; (R) plate-like or granular; (B) spherical or cubic. Both (R) and (B) can be described as crystalline.

Dates in Use: Since early 20th century

Density: (Y) 3.5–4.3g/cm³; (R) 4.8–5.2g/cm³; (B) 4.6–5.2g/cm³

Hardness: (Y) 5–5.5; (R) 5.5–6.5; (B) 5.5–6 on Mohs scale

Oil Absorption Ratio: (Y), (R), (B) 15–25g of oil per 100g of pigment

Refractive Index: (Y) 2.2–2.4; (R) 2.9–3.2; (B) 2.42

Health and Safety: Generally considered non-toxic, but prolonged exposure to dust may cause respiratory irritation

Environmental Impact: Raw material extraction of iron ore present environmental issues; but non-toxic pigments pose little threat when disposed of

Left: A selection of iron oxide pigments made from calcining a single sample of iron oxide, gifted from Cornish Metals Inc. (South Crofty Mine) in Cornwall, in collaboration with The Royal Cornwall Museum.

THE EFFECTS OF CALCINATION

In Chapter 1 we saw how calcination was understood by our prehistoric ancestors as a means to change the colour of ochre. Iron oxide undergoes significant transformations when subjected to calcination at varying temperatures, affecting its phase, particle size, shape, and structure – and consequently its use in paint formulations. Let's start by looking at the changes in pigment properties that occur during calcination

CHANGES IN PIGMENT PROPERTIES DURING CALCINATION

PARTICLE SIZE AND SINTERING

Calcination causes small iron-oxide particles to coalesce into larger ones through the sintering process. As particles heat up, they lose some of their surface energy, which allows them to merge into bigger aggregates.

Effect: Larger particles have less surface area per unit of volume, reducing their capacity to scatter light. This decrease in surface area can result in reduced opacity and tinting strength in paints. Larger particles also affect the texture and consistency of the paint, making it less smooth and possibly more challenging to apply evenly.

CRYSTALLINITY

Calcination typically enhances the crystalline structure of iron oxides, especially when temperatures are high enough to promote crystal growth. For instance, amorphous or less crystalline forms of iron oxide, such as goethite, are converted to more crystalline forms like hematite.

Effect: Higher crystallinity generally increases the pigment's stability and durability, making it more resistant to weathering, UV light, and chemical degradation. However, it also makes the pigment less reactive, which can reduce its dispersibility in paint formulations. Poor dispersion can lead to clumping or uneven colour distribution.

Below: Two shades of iron oxide gifted from Cornish Metals Inc. (South Crofty Mine) in Cornwall, in collaboration with The Royal Cornwall Museum. Orange is unburnt/raw and brown has been calcined to 900°C.

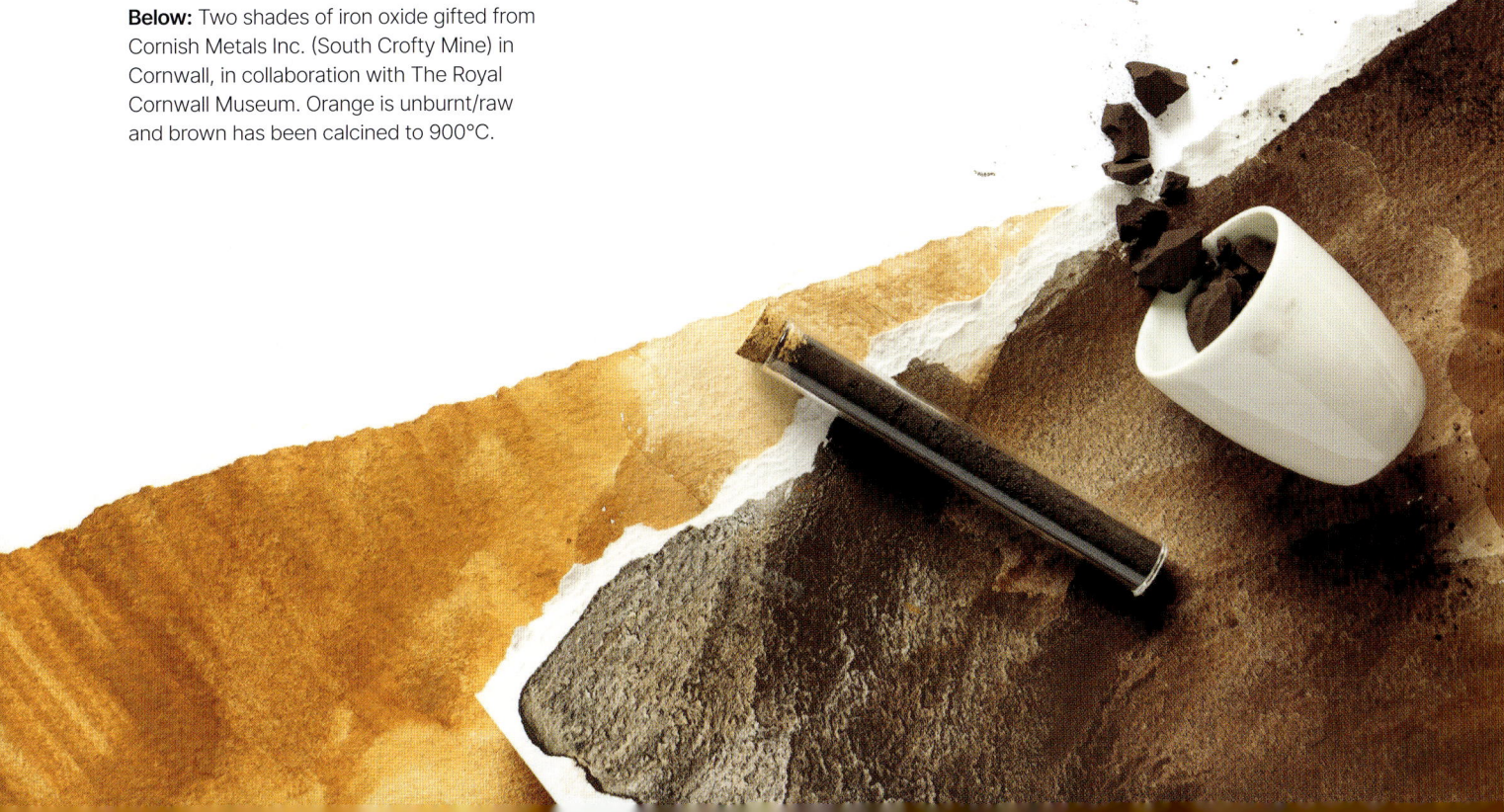

COLOUR INTENSITY

The colour of iron oxide pigments changes during calcination, particularly when yellow ochre (goethite) is heated to produce red ochre (hematite). The increase in temperature dehydrates the iron hydroxide, intensifying the red hue.

Effect: The more intense red colour of calcined iron oxide is prized for certain paint applications where strong colouration is required. The hue may become less vibrant in high-temperature calcined pigments due to larger particle size, which scatters less light and dulls the colour.

OPACITY AND TINTING STRENGTH

As particles grow during calcination, the ability of the pigment to scatter light diminishes, reducing the opacity of the paint. This can be a disadvantage in applications where strong coverage is required, forcing the painter to apply more layers to achieve the desired effect.

Effect: The reduced opacity affects the paint's ability to mask underlying surfaces, making the pigment less efficient in terms of hiding power. Tinting strength is also compromised, meaning more pigment must be used to achieve the same colour saturation.

Below: Natural minerals (sourced from Clearwell Caves, Gloucestershire) alongside their synthetic pigment counterparts

SURFACE AREA AND DISPERSIBILITY

Calcination reduces the surface area of iron oxide pigments as particles fuse and grow. Lower surface area decreases the pigment's ability to disperse evenly in the paint medium.

Effect: Poor dispersion can lead to uneven colouration, streaks or clumping in the final paint film. Wetting agents or dispersants are often required to overcome these challenges. Additionally, a less dispersible pigment can increase the time and effort required during the paint-mixing process.

HARDNESS AND ABRASION RESISTANCE

Calcined pigments tend to be harder due to their crystalline structure. This can improve the pigment's resistance to mechanical abrasion and wear in industrial coatings, where durability is essential.

Effect: While increased hardness is beneficial for outdoor and protective paints, it may make the pigment more difficult to mill during paint production, requiring more energy and potentially causing wear on equipment.

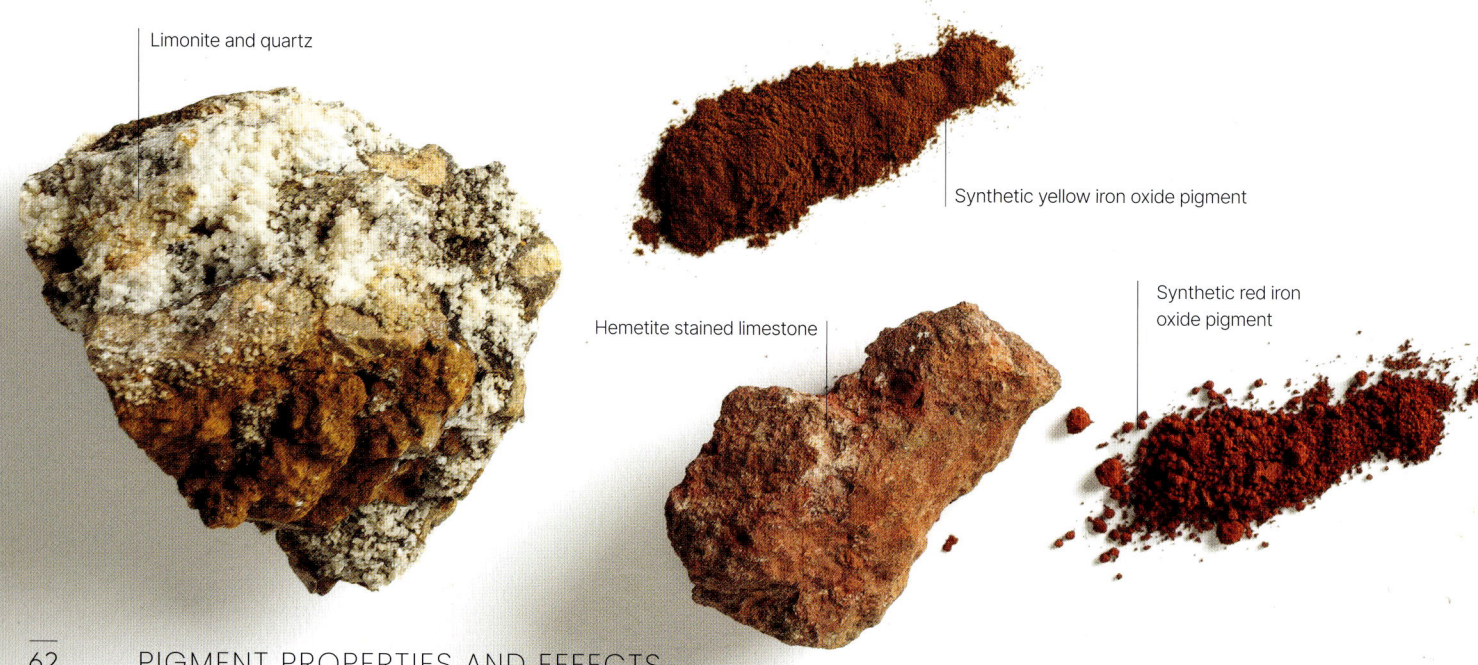

Limonite and quartz

Synthetic yellow iron oxide pigment

Hematite stained limestone

Synthetic red iron oxide pigment

THE SCIENCE OF TURNING YELLOW TO RED

Yellow ochre and red ochre are the common pigment names for goethite [α-FeO(OH)] and hematite [Fe_2O_3]. Both these minerals contain forms of iron oxide, and are found naturally occurring but rarely pure, often being present alongside silicate minerals such as clays, quartz and feldspars, calcium compounds, other metal oxides, and carbonates, etc.

Goethite is an iron oxide hydroxide – a hydrated iron oxide, containing water in its makeup. Hematite is a common iron oxide mineral, in its purest form containing 70% iron and 30% oxygen. Under certain conditions both compounds are formed as common weathering products of iron-bearing minerals like magnetite and pyrite. Hematite is the primary source of iron metal for the steel industry. Limonite was the general name given to weathering products, before modern mineral analysis, and is still used today to describe yellow or brownish iron oxides.

We can create red ochre synthetically. On heating, goethite can be burnt (calcined) into hematite. This change is due to the dehydration of the material; through a loss of water content, the mass and structure of the particles change. Goethite dehydrates at temperatures in the range 250°C (483°F) up to approximately 700°C (1292°F) to form hematite. The longer the exposure to heat and the higher the temperature, the darker the resulting red.

In nature, this transformation occurs where geological processes, such as burial or volcanic activity, create the iron oxide state conversion.

It is not unusual for these pigments to look very different in their various forms. When found as minerals in nature, or ground into a fine powder, they might look darker or more dull. It is in the processing of the raw material into a finely ground powder and its dispersion in a binding medium that its colour is truly revealed. A test for their colour potential is this: when rubbed on a raw or bisque-fired ceramic surface, hematite leaves a reddish-brown streak, while goethite leaves an ochre-yellow colour. This is known as a mineral's 'streak mark', and aids the confirmation of a suspected mineral type by geologists. Alternatively, use a piece of 240-grit sandpaper to do this test.

Particle size, shape, and crystal structure can all affect the resulting hue of the pigment. These qualities help determine the handling of the paint, its transparency, and how fast it dries.

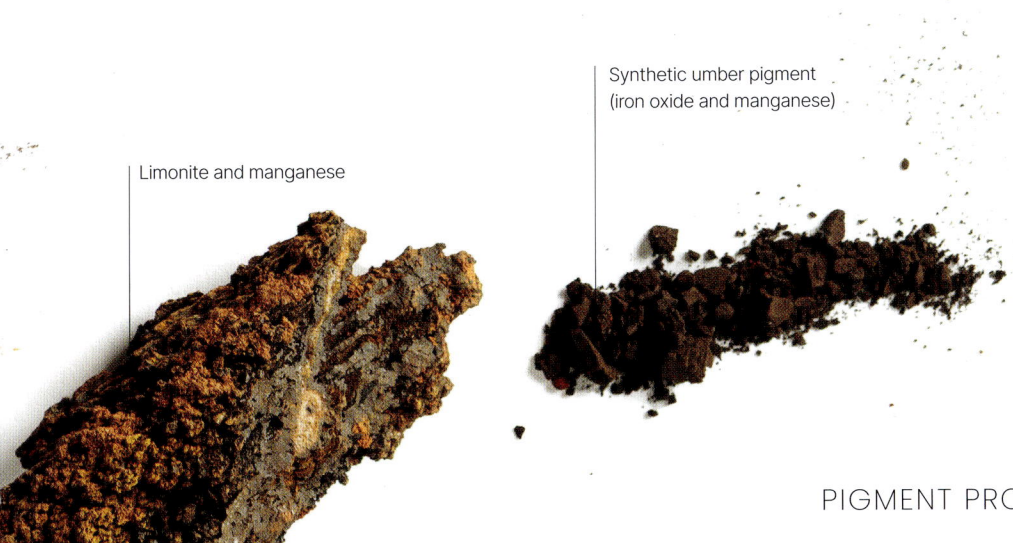

Synthetic umber pigment
(iron oxide and manganese)

Limonite and manganese

CHANGES IN STATE OF IRON OXIDE DURING CALCINATION

The following exploration of the chemical and physical changes to iron oxide can be explored through the recipes for calcination (see Recipe: Calcination in Chapter 1), with any colour changes observed to be explained here and in the following sections. At pivotal temperatures (and sometimes under particular conditions and with the addition of other chemicals), iron oxide particles go through unmistakable and prominent changes. These changes are referred to as phases.

STATES OF IRON OXIDE

A phase in the context of iron oxide refers to a specific form or state of the material, which can have a distinct crystal structure, chemical composition and physical properties. Iron oxides exist in various phases, depending on the oxidation state of the iron (Fe) and the arrangement of oxygen atoms. Each phase has unique characteristics that determine its colour, stability, and behaviour in different applications, including as pigments.

The iron oxide starting material for calcining at different temperatures is hydrated iron oxide. As a commercial artist's pigment it is sold as 'synthetic yellow oxide' or naturally occurring 'yellow ochre', which will also contain other minerals in the mix. To observe the changes outlined in this section accurately, it is advised to purchase the purest grade of hydrated iron oxide, or carefully process your yellow ochre to remove impurities using levigation (see Processing Raw Earth Into Your Own Pigments in Chapter 6).

Temperatures are approximate owing to the unknown constituent parts of your handmade or store bought yellow ochre alongside the use of non-laboratory-grade equipment.

COMMON PHASES OF IRON OXIDE

Goethite (α-FeO(OH)) – A yellow-brown hydrated iron oxide, found in natural earth pigments.

Lepidocrocite (γ-FeO(OH)) – A metastable orange-red form, often associated with rust.

Hematite (α-Fe$_2$O$_3$) – A stable, reddish-brown phase, commonly found in rust and natural red ochres.

Magnetite (Fe$_3$O$_4$) – A black, magnetic phase containing both Fe^{2+} and Fe^{3+}.

Maghemite (γ-Fe$_2$O$_3$) – A metastable, magnetic brownish-red phase, structurally similar to magnetite but partially oxidized.

Scientific notation explained:

α – *represents the stable or dominant phase.*

γ – *represents a metastable or alternate phase that may be less stable or form under specific conditions.*

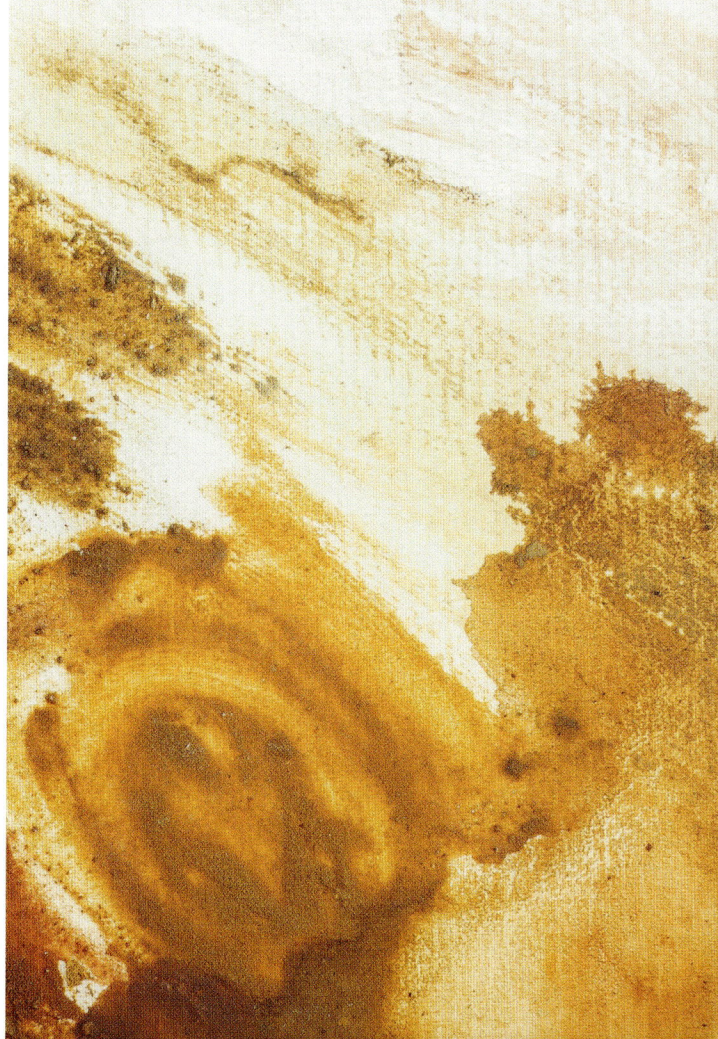

STARTING MATERIAL

Pigment, synthetic or natural that is rich in the hydroxides of iron: Goethite (α-FeO(OH)), Lepidocrocite (γ-FeO(OH))

DEHYDRATION

TEMPERATURE: Below 200°C (392°F)

Phases of iron oxide material: Goethite (α-FeO(OH)), Lepidocrocite (γ-FeO(OH))

REACTION: At low temperatures dehydration occurs. Hydroxides of iron that contain water in their particle structure will have the water driven out through evaporation, leaving behind a less hydrated or anhydrous form of iron oxide.

Particle characteristics: Begin to shrink as water is lost, leading to a change in the surface structure.

Phases of iron oxide formed or present: Goethite (α-FeO(OH)) $\rightarrow$ dehydrates to $\rightarrow$ Hematite (α-Fe$_2$O$_3$)

Phases that may still be present: Goethite (α-FeO(OH)), Lepidocrocite (γ-FeO(OH))

LOW-TEMPERATURE CALCINATION

TEMPERATURE: Around 200°C to 500°C (392°F to 932°F)

Phases of iron oxide material: Goethite (α-FeO(OH)), Lepidocrocite (γ-FeO(OH))

REACTION: As the temperature increases, hydrated iron oxides are converted into anhydrous iron oxide (hematite), the most stable form of iron oxide at higher temperatures. The dehydration reactions produce hematite through the loss of hydroxyl groups characterized by its reddish-brown colour and dense crystalline structure.

Particle characteristics: Hematite particles formed at lower temperatures tend to be smaller, irregular, and have a rough surface.

Phases of iron oxide formed: Lepidocrocite (γ-FeO(OH)) $\rightarrow$ dehydrates to $\rightarrow$ Hematite (α-Fe$_2$O$_3$)

Phases that may still be present: Goethite (α-FeO(OH)), Lepidocrocite (γ-FeO(OH)), Ferrihydrite (Fe$_5$HO$_8 \cdot$4H$_2$O), Hematite (α-Fe$_2$O$_3$)

INTERMEDIATE-TEMPERATURE CALCINATION

TEMPERATURE: 500°C to 800°C (932°F to 1472°F)

REACTION: At these levels any remaining goethite or other iron oxyhydroxides fully transform into hematite. Additionally, under specific conditions, particularly in a slightly reducing atmosphere or with incomplete oxidation, magnetite can form instead of hematite. Magnetite is a black, magnetic iron oxide phase. Maghemite, a metastable phase, can form through partial oxidation of magnetite. Maghemite is often a transition phase before converting to hematite upon further heating.

Particle characteristics: The hematite particles start to grow in size as sintering (particles fusing together at high temperatures) occurs. The shape becomes more defined, often crystalline, but the size remains moderate. The particles have smoother surfaces but maintain good pigmentary qualities.

Phases of iron oxide formed or present: Magnetite (Fe$_3$O$_4$) $\rightarrow$ oxidizes to $\rightarrow$ Maghemite (γ-Fe$_2$O$_3$) $\rightarrow$ further oxidizes to $\rightarrow$ Hematite (α-Fe$_2$O$_3$)

HIGH-TEMPERATURE CALCINATION

TEMPERATURE: 800°C to 1000°C (1472°F to 1832°F) and above

REACTION: At high temperatures, iron oxide particles continue to sinter, leading to the formation of more uniform, larger, denser particles. Hematite remains stable up to these temperatures, but its crystalline structure becomes more refined. High-temperature calcination promotes the full conversion to hematite if any other forms of iron oxide are present. In extremely reducing conditions at high temperatures, wüstite can form. This phase is unstable in the presence of oxygen and will easily convert to other forms of iron oxide upon cooling or in an oxidizing environment.

Particle characteristics: The iron oxide particles grow larger due to sintering, with smoother, more regular shapes. The increased particle size and density can lead to a loss in pigment opacity and a reduction in tinting strength.

Phases of iron oxide formed or present: Wüstite (FeO) $\rightarrow$ oxidizes or heats to $\rightarrow$ Hematite (α-Fe$_2$O$_3$) (in oxidizing conditions)

HANDLING IN PAINT FORMULATION

As calcined iron oxide pigments become more crystalline and harder, they can be more difficult to break down and evenly disperse in the paint medium. The increased particle size and decreased surface area require stronger mechanical action during the mixing and milling process. Paint manufacturers may need to use higher-energy mills or additional additives to achieve a uniform dispersion.

Paints made with calcined pigments may exhibit changes in texture and flow. Larger, more crystalline particles can result in a rougher texture and affect the paint's ability to form a smooth, even coat. This can be a concern for applications requiring fine details or smooth surfaces.

Due to their increased stability, calcined pigments contribute to the formation of more durable paint films, making them ideal for exterior applications. However, the loss of opacity and tinting strength may require adjustments in the formulation, such as increasing pigment load or combining the calcined iron oxide with other pigments to achieve the desired balance of colour and coverage.

SPECIFIC USES OF CALCINED PIGMENTS IN PAINTS

INDUSTRIAL AND PROTECTIVE COATINGS

The high stability, hardness, and durability of calcined iron oxide pigments make them suitable for use in protective coatings for metals, concrete, and other surfaces exposed to harsh environments. The pigments' resistance to UV light and chemicals ensures that the coatings maintain their protective properties over time.

ARCHITECTURAL PAINTS

Calcined pigments may be used for both aesthetic purposes and protection. While the reduced opacity may require more layers, the stability and colourfastness of the pigment ensure that the paint will not degrade quickly under sunlight or weathering.

ARTISTIC AND DECORATIVE PAINTS

For fine-art applications, the colour vibrancy of calcined pigments may be desirable, though their handling requires careful formulation to ensure proper dispersibility and smooth application.

Right: Calcining a sample of iron oxide in a crucible with an enameling kiln.

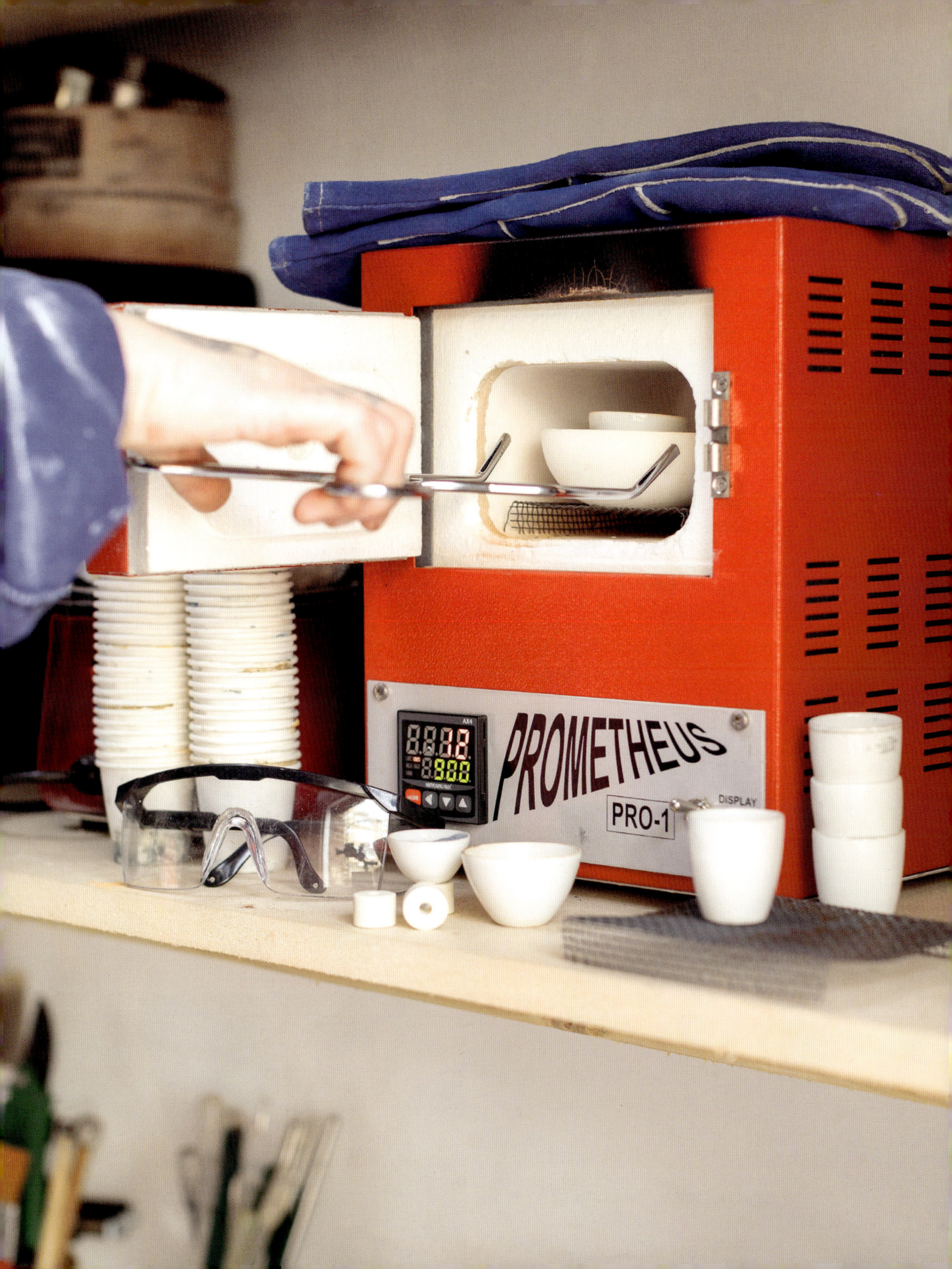

PAINT APPLICATIONS BY IRON OXIDE PHASE

Different phases of iron oxide occur depending on the temperature, pressure, and surrounding atmosphere during calcination.

CRYSTALLINE STRUCTURE: A phase describes how the atoms in the material are arranged. For example, iron oxide can crystallize into different structures like hematite (α-Fe_2O_3), magnetite (Fe_3O_4), and maghemite (γ-Fe_2O_3). These phases have different spatial arrangements of the iron (Fe) and oxygen (O) atoms, which give them distinct properties.

CHEMICAL COMPOSITION: Some phases of iron oxide have slightly different chemical formulas. For example, hematite (α-Fe_2O_3) is composed purely of ferric iron (Fe^{3+}), whereas magnetite (Fe_3O_4) contains both ferric (Fe^{3+}) and ferrous (Fe^{2+}) iron ions.

THERMODYNAMIC STABILITY: The phase of iron oxide that forms at a given temperature and atmosphere depends on its thermodynamic stability under those conditions. For example, hematite is stable at high temperatures in an oxygen-rich environment, whereas wüstite (FeO) is stable only in a highly reducing environment at elevated temperatures.

PHASE TRANSITIONS: As iron oxide is heated during calcination, it can undergo phase transitions, where it changes from one phase to another due to changes in temperature or other conditions. For example, hydrated iron oxides such as goethite (FeO(OH)) lose water and convert into hematite (Fe_2O_3) as the temperature increases.

Understanding the phases of iron oxide during calcination is crucial for tailoring the properties of the material to specific uses, such as pigments in paints or industrial coatings. See Common Phases of Iron Oxide for a handy summary.

APPLICATIONS OF IRON OXIDE PHASES

Name: Goethite (α-FeO(OH))

Colour: Yellow to Brown

Uses: Earth tone pigments in art paints, exterior coatings

Name: Hematite (α-Fe_2O_3)

Colour: Red

Uses: Art paints, anti-corrosion primers, industrial coatings

---

Name: Magnetite (Fe_3O_4)

Colour: Black

Uses: Magnetic paints, anti-corrosion coatings, industrial use

---

Name: Maghemite (γ-Fe_2O_3)

Colour: Reddish-brown

Uses: Magnetic paints, industrial and art paints

Name: Wüstite (FeO)

Colour: Black (unstable)

Uses: Rare specialized coatings in controlled environments

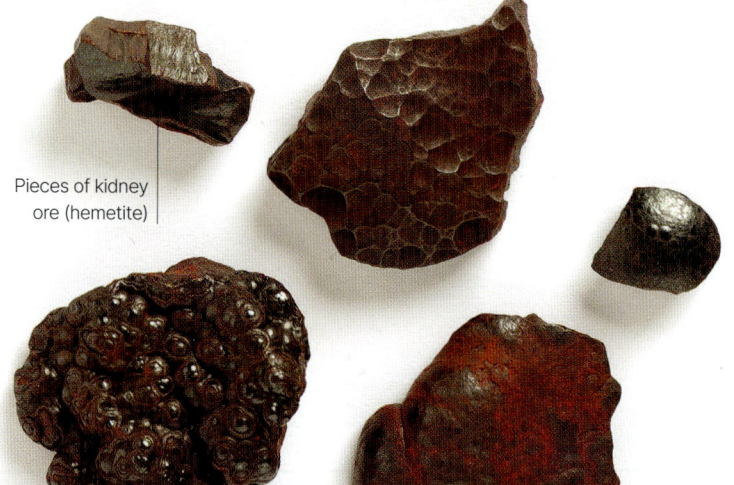

Pieces of kidney ore (hemetite)

EARTH PIGMENTS WITH A PURPLE SHADE

Purple iron oxide pigments can be found both naturally and synthetically. In natural purple ochres, iron oxide is the primary colouring component, while manganese in the earth further shifts the hue towards purple. However, intensely purple ochres are rare, as the fine grinding required for use as pigments often shifts their colour to red. Moreover, these pigments are typically mixed with other minerals in their natural deposits, further affecting their purity and intensity.

NATURALLY OCCURRING PURPLE EARTHS

Natural purple earths form through geological processes involving iron and manganese oxides. Their colour results from a combination of iron oxide (hematite or goethite) and varying amounts of manganese compounds. These earths typically develop in sedimentary and weathered metamorphic environments under specific conditions. Weathered hematite often develops purple tones, which result from the formation of large iron oxide particles on the surface of the mineral.

In *A General Natural History* (1748), John Hill classified various types of native mineral purples by colour, mass, and texture, linking them to ancient pigment names like sil, sinopis, and rubrica. In 1841 George Field describes the pigment as 'murrey' or chocolate-coloured, forming cool purple tints with white. Indeed, native mineral purples have long been found in the Forest of Dean, southern England, from the Clearwell Caves.

COLOUR VARIATIONS: The appearance of purple is due to how light interacts with larger, unground particles as they scatter light in a way that gives a cooler, more bluish or purplish hue. When the pigment is finely ground, the particle size decreases, altering the way light is absorbed and reflected. This often results in a shift towards a warmer, redder appearance, as the finer particles absorb more blue light and reflect more red wavelengths. Particles are measured in μm or microns (a micron is one-millionth of a metre): larger particles (1–5 μm) display a blue-red to purple hue, with smaller particles (0.1–0.2 μm) appearing reddish.

SYNTHETIC PURPLE OXIDES

Mars violet belongs to the family of Mars pigments, a group of synthetic inorganic pigments including Mars red that were developed in the 18th century. These are produced through the aqueous precipitation of iron salts (e.g., sulfates, chlorides, nitrates, and acetates) with an alkali, such as lime, caustic soda, or potash. This process initially creates yellow iron oxide hydroxide (Mars yellow), which is then roasted and oxidized to form red shades of iron(III) oxide.

CALCINATION: Mars violet can be created by calcining hematite or synthetic iron oxide. This process involves heating iron oxide at high temperatures, causing a shift in hue to violet-brown tones as larger particles form. These colours are marketed as 'Mars Purple' or 'Mars Violet' and 'Caput Mortuum' in commercial settings, with the latter being a closely related pigment. Many 19th-century pigment manufacturers, including London's artist's colourman C. Roberson & Co. (established in 1810) enhanced Mars violet pigments by adding an organic violet dye, intensifying their chromatic vibrancy.

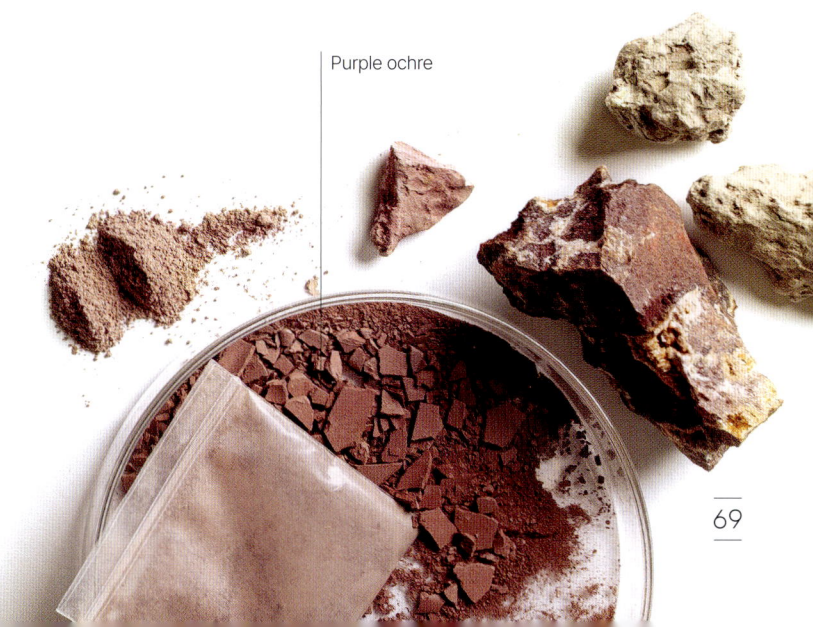

Purple ochre

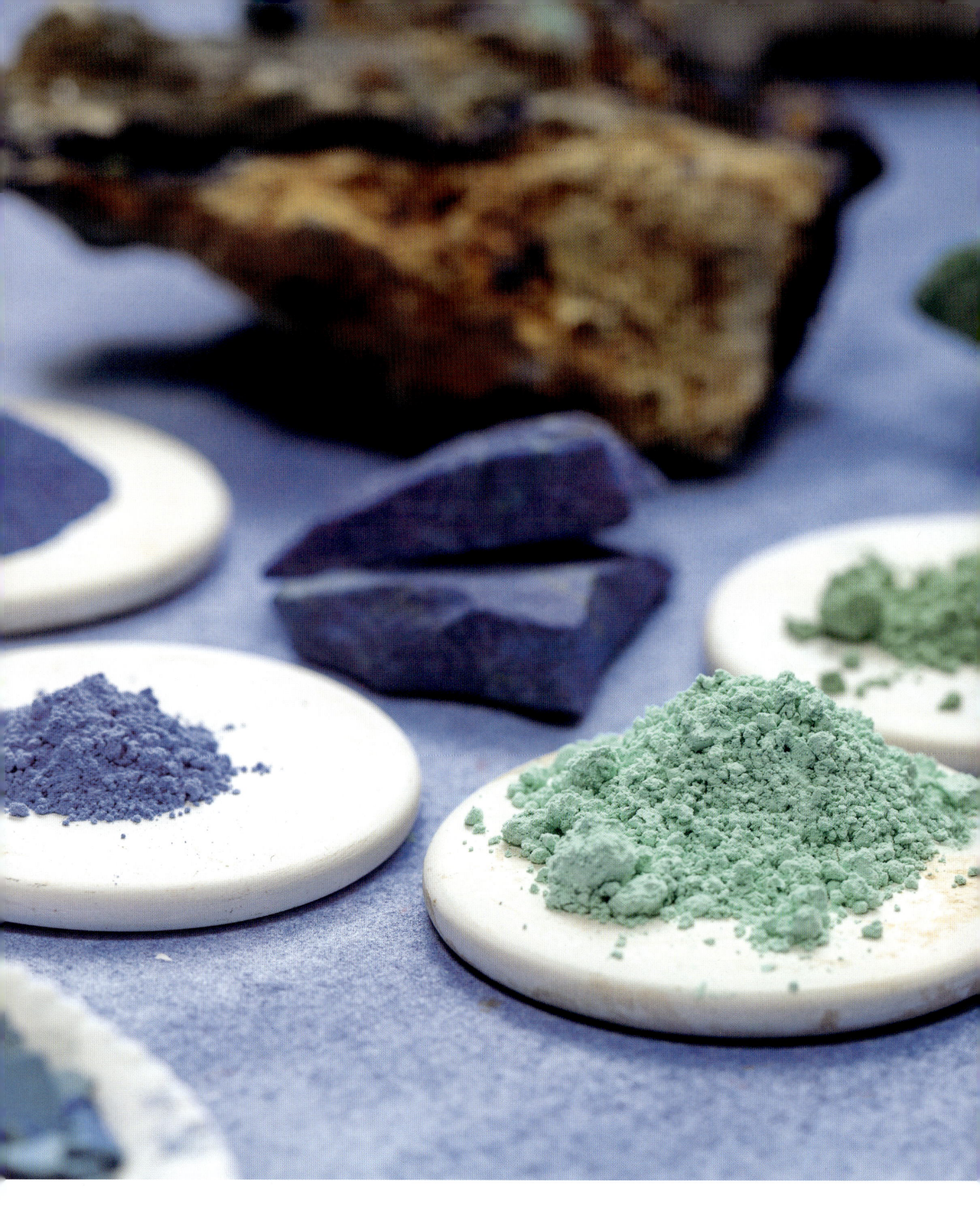

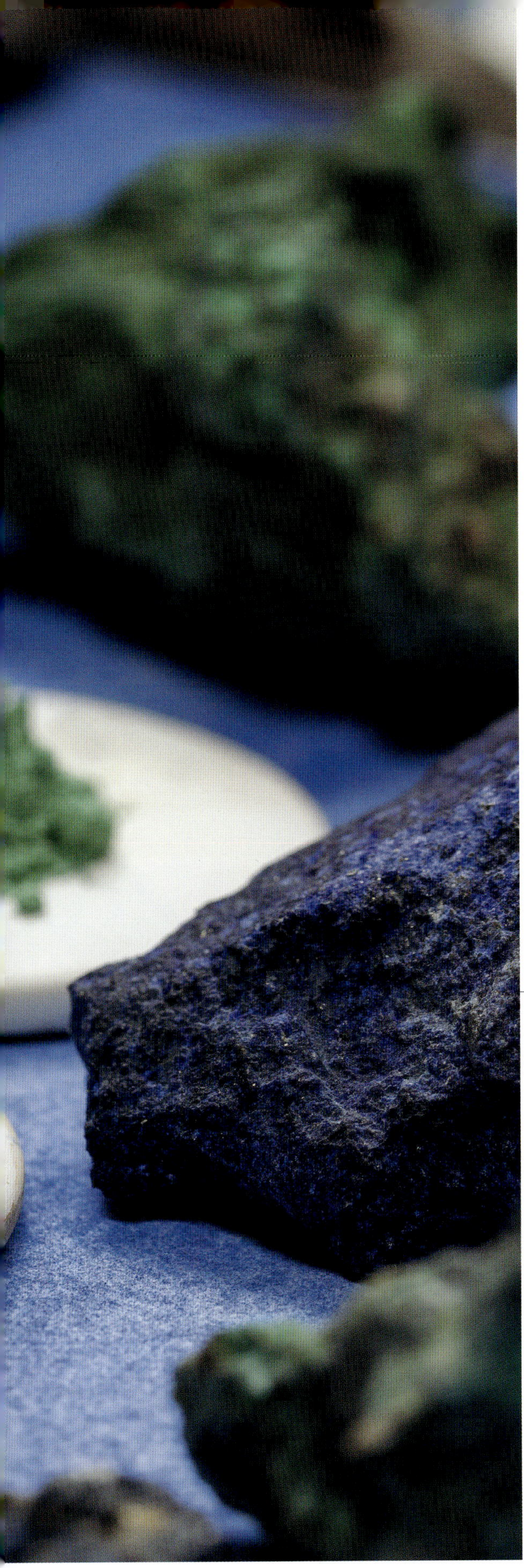

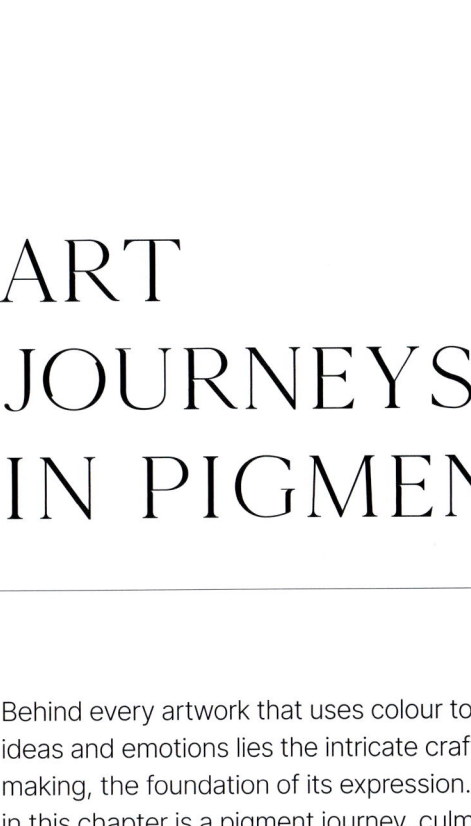

CHAPTER FOUR

ART
JOURNEYS
IN PIGMENT

Behind every artwork that uses colour to convey ideas and emotions lies the intricate craft of pigment making, the foundation of its expression. Each story in this chapter is a pigment journey, culminating in a recipe for you to continue the story in your own way. The chapter considers a small selection of pigment-making traditions and technologies, and the works of art associated with them. We look at how pigment making has affected visual culture. The pigments highlighted here are lapis lazuli and its human-made counterpart, synthetic ultramarine, as well as azurite and malachite.

TRADITIONS AND TECHNOLOGIES

The word 'technology' may suggest great leaps in science or industry, but its roots are rather different. It comes from the ancient Greek *tékhnē* ('art' or 'craft'), while *logia* connotes 'discourses' – collections of words, written down or orally transmitted. We can consider the technologies that changed the way we think rather than automated what we could already do by hand: hand-based technologies, tactile technologies, craft technologies. By learning about traditional handcrafts we learn more about ourselves, through connecting with our more-than-human world. Blue and green pigments have been used widely for around 5,000 years, yet pigment usage goes back much further as we saw in Chapter 1. Archaeologists tend to agree that our species and other hominins (e.g Neanderthals) have been using rocks, stones, and minerals for paint for at least 500,000 years.

Earth has been our means to communicate and grow our ideas; it sustains us physically and mentally. The construction of the 'picture plane' in Western art – an imaginary space on a flat surface – and its evolution over time developed due to our ingenuity in pigment use, pushing boundaries in pictorial representation. Throughout time, every rock crushed, every sap collected, each dye extracted and new mark drawn led us to new ways to share stories and the complexities of our lives. In turn this meant we knew our surroundings intimately, every bird, moss, and stone (known in common parlance as 'the forager's eye').

By understanding ways in which we enriched ourselves through inquisitiveness and curiosity, we can perhaps find our way back to these innate human behaviours. Creativity is sometimes seen more as a skill and less of a fundamental human behaviour. This disconnect has separated us from our human capacity to engage with the natural world. Artistic skills have been commodified and given financial value, derailing our creativity and creating boundaries to its therapeutic qualities.

Art communicates ideas and emotions. This ability to communicate has been dictated not only by the timeline of our cognitive development but by the availability of materials and the processes and technologies needed to realize these ideas. But perhaps it's not always been this way round. The use of ochre and other mineral pigments has been linked by archaeologists to our brain development. Geophagy – literally eating earth, usually for nutrients – may have contributed towards brain development and could have led us to interact with ochre to create sophisticated value associations. We have long engaged with this material, even using it as an additive in adhesives for stone tools. Ochre can be regarded as a multifunctional, anthropomorphized substance.

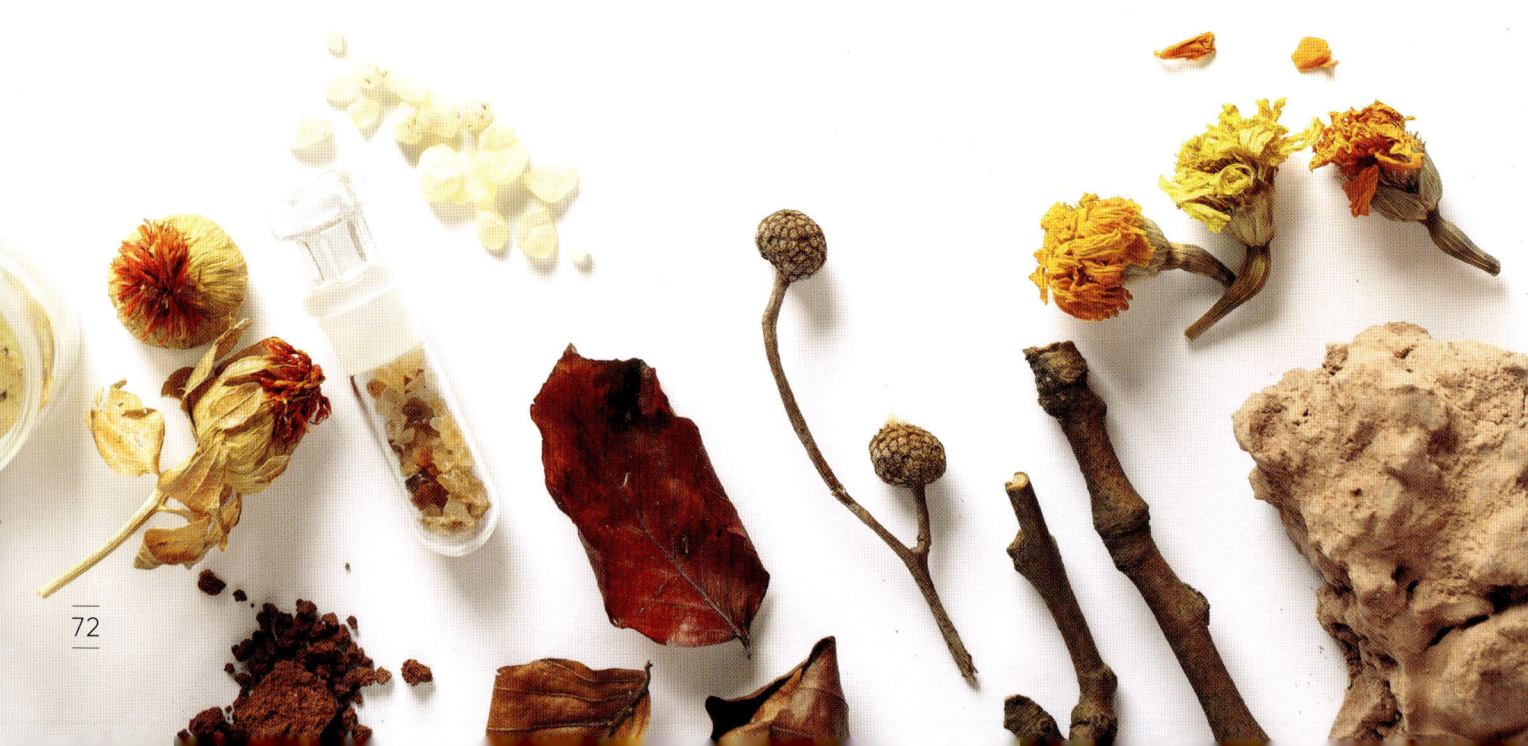

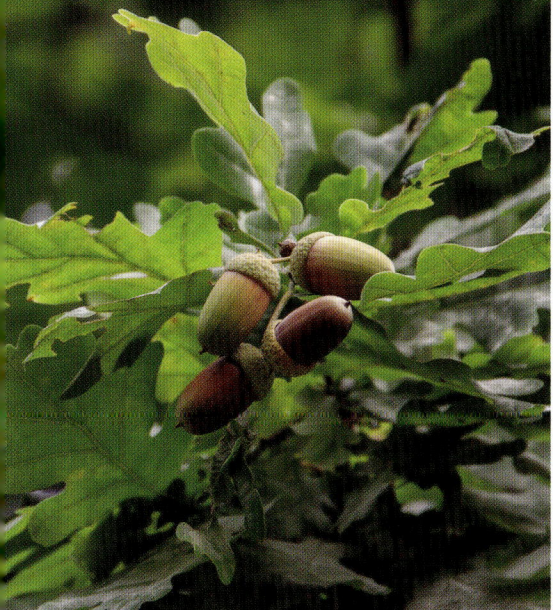

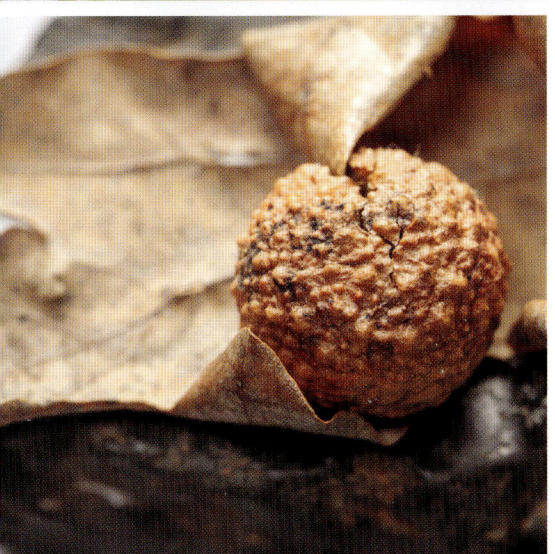

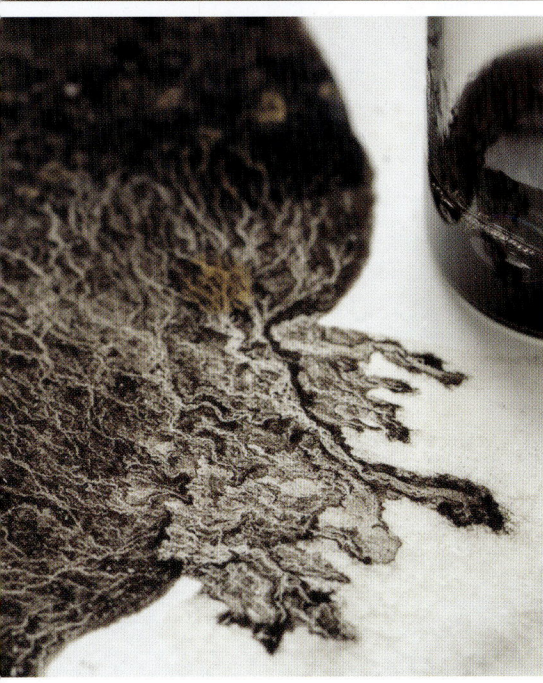

AN ARTIST'S ASIDE

REFRAMING OUR RELATIONSHIP WITH PIGMENTS

I was inspired by a talk given at Kew Gardens in May 2024 by Dr Robin Kimmerer, a Potawatomi botanist and director of the Center for Native Peoples and the Environment (based in Syracuse, New York State), and author of the bestselling book, *Braiding Sweetgrass* (2013). Western science has created boundaries to organize, stratify and break complex systems down in order to understand them. Through this way of thinking, we have objectified nature and created false boundaries. Although it is a part of us, we have 'othered' it, denied it. She calls for a kind of radical empathy where we respect all beings equally.

This call to action, or call to attention, is how I now look at pigments. Pigments are the lens through which I see the world, noticing how vibrant or dull the colour of earth is – knowing the different oak species, for example. Seeing the world for colour potential is paradoxically and concurrently both hyper-specific and general.

Kimmerer calls for the use of the word 'Ki' for individuals and 'Kin' for our collective family of organisms that includes everyone and everything. This resonates with the teachings of St Francis of Assisi in the *Canticle of the Creatures* (1223) and his discovery of 'created elements' as members of the same family: Sister Sun, Brother Moon, Mother Earth, Sister Water – a real sense of interconnectedness and reciprocity with the natural world. Or as philosopher Martin Buber put it in *Ich und Du* (*I and Thou*) (1923), we need to make the shift from an I-It relationship with nature to an I-Thou relationship – the object we want to manipulate becomes a subject we must listen to. This is a hugely transformative approach, turning upside-down and inside-out our usual, default utilitarian approach to physical elements.

Such approaches stimulate us to revisit our very view of pigments. So then, what are pigments? They are messengers, teachers, ecosystems, and living organisms. Pigments are animate – they move and they are tools for accessing and building soulful practices. They are part of us and we are part of them – we are a medium for nature.

JOURNEYING WITH ULTRAMARINE BLUE

My life with pigments has been a fascinating, meandering journey, whereby certain colours or techniques have brought me to the most surprising of places. Living for a year in Jerusalem, for instance, I encountered the Church of the Holy Sepulchre, known locally as the Church of the Resurrection. Via incense-laden Ethiopian chapels, one enters the basilica to navigate an ecclesiastical warren, where lapis lazuli flashes through the gloom.

I first fell under the spell of ultramarine blue – or rather lapis lazuli, its naturally occurring partner – through my father, an Anglican priest, now retired, and a Franciscan. From an early age, I enjoyed learning about Christian iconography, and revelled in the ritual and mysticism of my upbringing. Although I am agnostic, I have a great fondness for orthodox symbology and wall paintings: I grew up with a diverse backdrop of Russian and Greek icons, as my dad is an avid collector of them.

Lapis lazuli is that otherworldly, celestial energy we see giving light from within itself, used in religious artwork to make the heavens tangible and to celebrate the sanctity of the Virgin Mary. The desire to create the purest of all blues spurred pigment makers across the centuries to experiment and push the boundaries of the technologies they had. For the Renaissance painters, it was light. Fine pure lapis pigment was suspended in thin glazes of a translucent mastic painting medium. Layers built up over time upon the picture plane created physical space, allowing light to refract within the paint layer, bouncing around within it – effectively lighting the painting from the inside.

This technical knowledge can be seen in many museums across the world. One famous example, in London's National Gallery, is the painting known as *The Wilton Diptych*. One of the earliest known pieces of Western art using lapis, it was created in

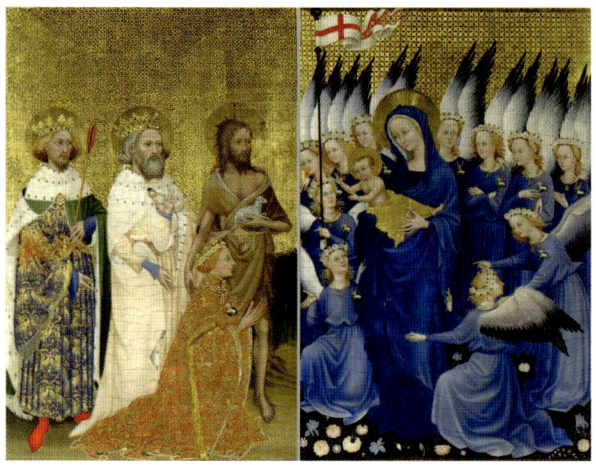

Above: *The Wilton Diptych*, (c. 1395–1399). Egg tempera on oak panel. The National Gallery, London.

c.1395–99. On one of the panels ten angels surround the Virgin Mary and Child in a vivid colourfield of lapis that takes up about 80% of the image. When I saw this work on my first visit to London when I was 15 years old, I remember consuming the blue and its opulent jewel-like colours with a hunger, and this reminds me now of the famous quote by novelist Ernest Hemingway, where he is literally looking at art with an empty stomach: 'The paintings were heightened and clearer and more beautiful if you were belly-empty, hollow-hungry. I learned to understand Cézanne much better and to see truly how he made landscapes when I was hungry.'

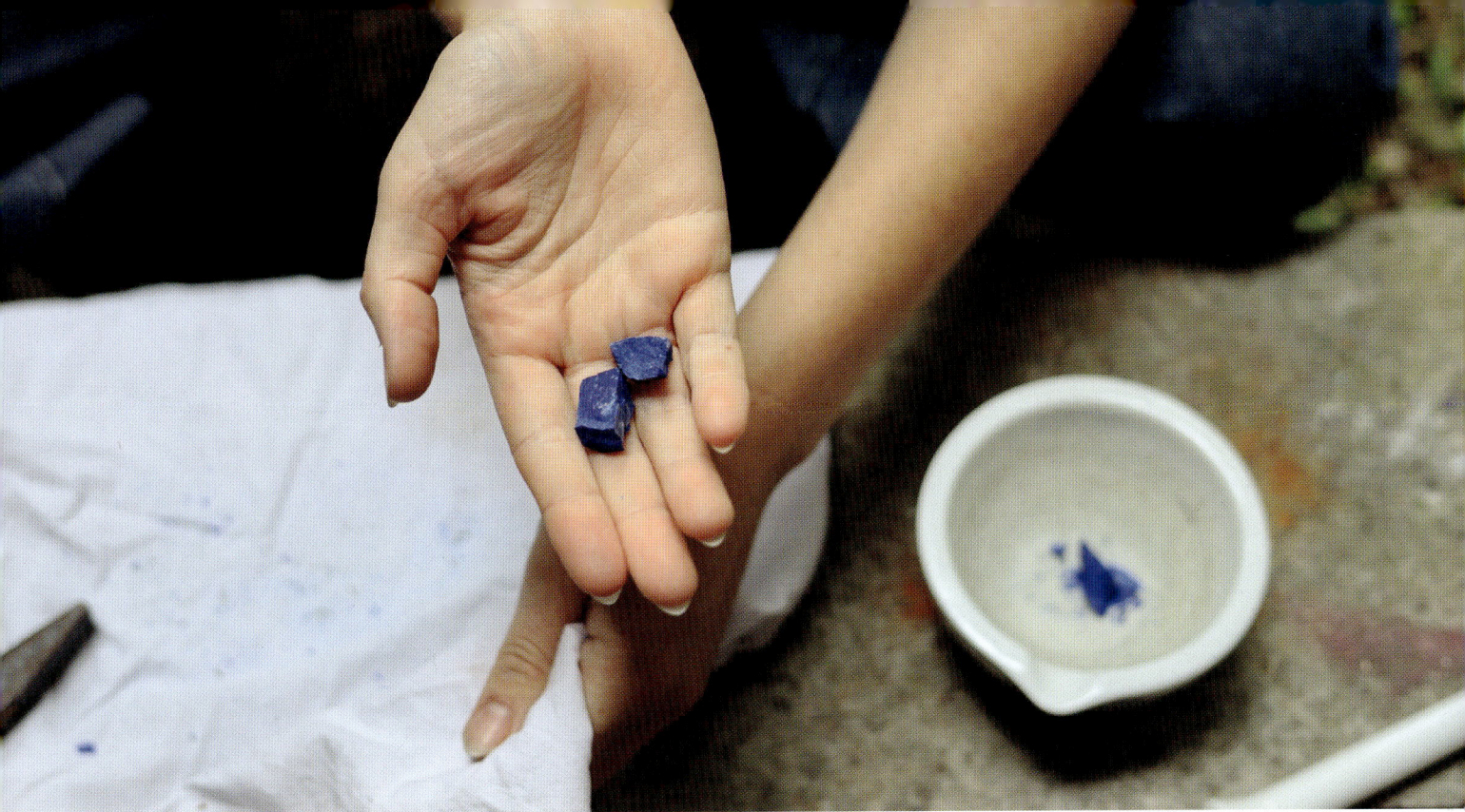

MAKING LAPIS LAZULI PIGMENT

In 2021, Jo Volley, artist, pigment aficionado and associate professor at the Slade School of Fine Art invited me, amongst other artists, to make some lapis lazuli pigment. For this experiment we followed the Fra Angelico recipe that Renaissance artist and biographer Cennino Cennini outlined in his book *The Craftsman's Handbook*.

Jo founded and runs the material research project at University College London and kindly supplied all the raw materials, some of which were gifted from the Winsor & Newton archive from Colart Group, the international supplier of art materials. The key ingredients were unbleached beeswax, mastic resin, colophony (also known as rosin), and finely ground lapis.

Mastic is a resin sourced from the mastic tree (*Pistacia lentiscus*), cultivated on the island of Chios, Greece, and also in parts of Turkey. The tree is expertly cut and a thick dusting of chalk is laid around its base, known as a 'table', to stop the fresh sap from sticking to the dirt. As the tears of sap fall

to the ground, clouds of bright white dust puff into the air. Like most traditional art materials, mastic has many useful properties and uses – from a food thickening agent to being an essential ingredient in chrism, the holy oil used for anointing in Eastern Orthodox churches and for the British monarch's coronation. It has medicinal and cosmetic uses, and it is added to coffee and sweet treats in Turkey and Greece. Dissolved in turpentine, mastic was used to create a picture varnish. However, it yellows over time and is comparatively soft in contrast to more hardwearing varnishes.

Colophony is the other resin used in the Fra Angelico concoction. Fresh sap from species of pine, mainly conifers, is heated to drive off solvents to extract the thick, viscous resin. As a solid, oleoresin, it is used in printing inks, varnishes, and soap, as well as to lubricate the bow when playing stringed instruments. As it is made from a wide variety of different pine species it can be harvested globally; however, the market leaders are the USA and Mexico.

RECIPE: LAPIS LAZULI

In this recipe, we follow similar proportions to those found in the 15th-century treatise *The Craftsman's Handbook* by Cennino Cennini, which details the methods and techniques of the masters.

INGREDIENTS

- Lapis lazuli stones, 1 part*
- Beeswax, 1 part
- Rosin, 1 part
- Mastic, 1 part
- Gum arabic, 1 part
- Lye (sodium hydroxide) solution, approximately 500ml per wash**
- A few drops of linseed oil

EQUIPMENT

- PPE – respirator (P3 rating), rubber gloves and safety goggles
- Hessian sack or reinforced plastic bag
- Metal mallet or hammer
- Magnifying glass
- Pestle and mortar
- Double boiler
- Washing-up bowl
- Wooden spoon or spatula
- Coffee filter papers
- Dehydrator (optional)

*A good starting quantity is 50g.
**Lye (also known as potash) can be substituted with an alkaline solution of either potassium carbonate or sodium carbonate instead, mixing 30g of either with 100ml of water.*

METHOD

CRUSHING THE STONES

1. Break the lapis lazuli into small pieces (pea size) with a hammer or mallet, working in a reinforced plastic bag or hessian sack to stop chunks flying around.

2. Use a magnifying glass to sort pieces into piles based on similar colours. Set aside any material containing impurities, such as white or grey fragments (calcite and quartz) or those with a metallic sheen or flecks (pyrite).

3. Selecting the bluest grades of stone pieces, grind them into a fine powder using a pestle and mortar. Set the ground lapis lazuli powder aside.

CREATING THE PUTTY

1. Melt the rosin, mastic, and wax together with the linseed oil in a double boiler. (You could also use a saucepan on a direct heat, but this will produce a lot of smoke.)

SAFETY NOTE

Lye solution is caustic; take every caution not to get it on your skin or in your eyes. Always wear gloves and a quality respirator or dust mask. Keep away from children and pets, and dispose of safely, diluting first with water before pouring down the drain.

2. Allow the mixture to cool slightly, so it becomes a malleable substance that can be handled rather than a liquid. Do this by testing its viscosity with a wooden spoon or spatula.

KNEADING THE PUTTY

1. Reheat the putty mixture in the double boiler and add the ground lapis lazuli powder, thoroughly mixing it in.

2. On a clean, smooth surface, knead the lapis putty, by pulling the dough and stretching it, before crimping it back together again. Keep kneading, rotating 90 degrees and repeating. It will be hard at first, but as it warms up it will become pliable. If it doesn't, put it into some boiling water to soften it.

3. Cennini calls for the putty to be kneaded over a period of three days so that the particles are evenly distributed. Repeat step 2 for at least 10 minutes per day.

WASHING THE PUTTY

1. Place the lye solution in a washing-up bowl. Massage the lapis putty in the lye solution. The lye will soften the resin mix allowing the lazurite particles to fall into the solution. The first release of particles will be the purest and brightest grade.

AN ARTIST'S ASIDE

THOUGHTS ON THE PUTTY METHOD

The teasing out of the blue is made possible by this exotic, alluring mix of symbolic and once rare and expensive materials. The blue is enriched with the smoke and perfume, anointed and manifested.

When I was making this mixture with Jo Volley, I wondered about all the fuss – I had made very bright blues from lapis stone before, by simply crushing and washing it in water. This works if the stone is particularly pure (and can often be an intense blue-purple, even black). However, the putty method, with its inclusion of the chrism ingredients used for anointing priests, kings, and candidates for baptism, emphasizes the association with liminality and transformation. This drawn-out process beautifies the colour, shrouding it in mystery and ritualizing its birth.

2. Once you feel the solution has turned a deep enough blue, leave the solution to sit and the pigment to settle. Pour the lye solution into a separate container, leaving the pigment behind. Dilute the lye solution with water and pour down the drain. Place your wet pigment in a coffee filter and wash with distilled water.

3. Repeat steps 1 and 2 until you have extinguished the colour from the putty, creating different grades of the pigment, from deep blue to a softer blue to a soft grey-blue.

DRYING AND GRINDING

1. Leave the resulting grades of pigment to dry on the filter papers (or dry in a dehydrator).

2. Package up your colours once dry, or experiment by grinding them further. There will be a moment during the particle reduction process, a sweet spot, where the pigment will be as fine as possible without desaturating. This is only learnt from experience.

PIGMENT SUMMARY

Pigment Name: Lapis lazuli

Nomenclature: Ultramarine, lapis lazuli, Fra Angelico blue, azure blue

Pigment Index Number: PB 29

Colour: Deep blue with purple hue

Chemical Name: Sodium aluminium silicate with sulfur

Chemical Formula: Lazurite $(Na,Ca)_8[(S,Cl,SO_4OH)_2(Al_6Si_6O_{24})]$

Classification: Natural inorganic

Description: Prized for its intense blue colour, it is composed primarily of the mineral lazurite, along with various other silicates

Conservation Issues: If not processed properly, impurities such as pyrite can lead to a dulling or yellowing of the vibrant blue over time; this can weaken the paint film, causing cracking and flaking, compromising the overall stability and longevity of the artwork

Particle Morphology: Crystalline, irregular grains with some fragments forming a granular texture when ground

Dates in Use: Ancient civilizations (particularly Egypt and Persia); Renaissance Europe

Density: 2.7–2.9 g/cm³

Hardness: 5–5.5 on Mohs scale

Oil Absorption Ratio: 40–60g of oil per 100g

Refractive Index: 1.50–1.60

Health and Safety: Non-toxic in its ground form, though inhalation of dust during grinding can be harmful

Environmental Impact: Lapis lazuli stone, primarily sourced from Afghanistan's Badakhshan region, raises significant ethical and environmental concerns

MAKING PAINT WITH YOUR LAPIS LAZULI

Simply choose your grade of pigment and mix it with your chosen binder using a pestle and mortar. Generally speaking, a good place to start is using a 1:1 ratio of binder to pigment. The proportion of pigment to binder will always vary due to the particle morphology: the size, shape, structure, and uniformity. Be careful not to grind your paint too much as it might reduce the particle size of the pigment, which will desaturate the colour.

DETERMINING IF LAPIS PIGMENT IS GENUINE

Pigment forgeries are rife – the spiking of colours with fillers or adulterants is one of the reasons I started making my own. To be sure that a lapis pigment is genuine, one method is to locate the sweet rank smell of sulfur. A sample is ground using a pestle and mortar, until the smell is emitted from the pigment. Sulfur is one of the main elements that makes up lazurite, alongside silica. Sulfur can be smelled during the manufacture of synthetic ultramarine too.

Mixing lapis lazuli pigment with water can also reveal hidden truths. If the pigment settles and the water is clear, there is no water-soluble adulterant. Whilst working at Cornelissen I spent several weeks during winter, in the outside yard at the back of the London shop, washing an impure batch of the pigment.

Ethically sourcing raw materials is essential for sustainability and integrity. When a company openly shares its supply chain and takes pride in supporting other businesses, this transparency enhances its authenticity. Prioritize grassroots, indigenous, and local retailers and manufacturers to promote fair trade and community growth.

RECIPE: SYNTHETIC ULTRAMARINE

Lazurite is a deep-blue mineral that is the primary component of lapis lazuli, a semi-precious gemstone. Synthetic ultramarine recipes replicate the formation of this mineral in situ in the kiln. Lazurite, a complex aluminosilicate of sulfur compounds, requires heating kaolin with sodium carbonate and sulfur. These materials are made or purchased in high purity concentrations, and come dry and powdered. Once mixed, charcoal is added as a catalyst in the reaction in many, but not all, synthetic. ultramarine recipes. Rosin, which melts at about 110°C (230°F) to form a fused, viscous mass, is added to stop the powders becoming airborne, damaging the kiln or combusting. Sulfur dioxide is created during the reaction. As this contributes towards acid rain pollution, commercial manufacture of ultramarine should include the safe handling of this by-product.

The shade of synthetic ultramarine varies significantly based on ingredient ratios, timing, temperature, and oxygen availability. In the sample (below), areas with optimal conditions show a more intense blue.

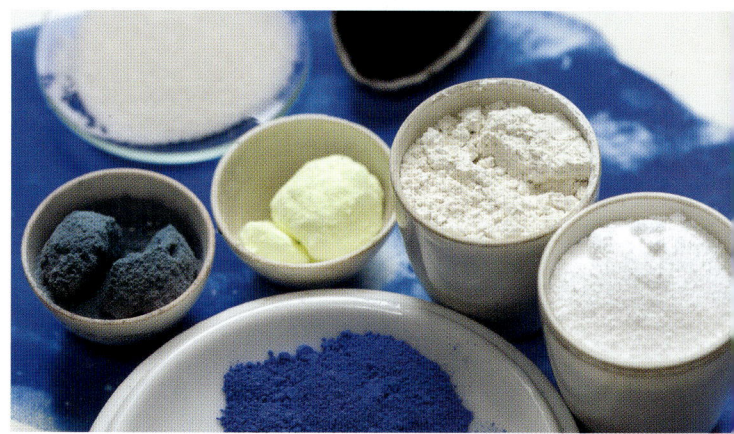

INGREDIENTS

- Kaolin, 100 parts
- Sodium carbonate, 100 parts
- Sulfur, 80 parts
- Charcoal, 12 parts
- Rosin, 40 parts
- Distilled water
- Acetic acid or white vinegar
- Natural surfactant of choice (see Natural Surfactants)

EQUIPMENT

- PPE – respirator (P3 rating), gloves and safety goggles
- Crucible and lid
- Kiln – I used an enamelling kiln (Prometheus Pro)
- Kiln shelf and wire rack*
- Inert spatula (plastic, stainless steel, glass)
- Pestle and mortar (laboratory-grade porcelain to prevent staining)
- Glass petri dishes
- Coffee filter papers
- pH indicator strips
- Dehydrator

*A kiln shelf protects the kiln from spillages; rest the crucible on a wire rack on removal from the kiln.

METHOD

PRETREATING THE KAOLIN

1. Calcine the kaolin separately by placing it in a crucible (without lid) in the kiln at 700°C (1292°F) for 1 hour and let cool. This can be done within an oxygen-rich atmosphere. This will make the particle shape amorphous, changing its size and shape to become a more reactive material known as metakaolin.

CREATING THE RAW PIGMENT

1. Mix the kaolin, sodium carbonate, sulfur (60 parts only), charcoal, and rosin using a pestle and mortar.

2. Place mixture into a crucible with a loose fitting lid, and place in the centre of the kiln. Bring kiln temperature (in a low oxygen environment) to 750°C (1382°F) for 4 hours.

3. Let the mixture cool, and tap out the raw pigment cake onto a clean surface. Inspect and collect the brightest blue parts into a dish for further processing. Dispose of grey or discoloured parts.

BRIGHTENING THE BLUE: SECOND CALCINATION

1. Grind the selected sample using a pestle and mortar with an additional 20 parts of sulfur.

2. Place in crucible (without lid)** and calcine with oxygen for 2 hours at 500°C (932°F).

3. Let cool slowly by lowering temperature incrementally by 50°C (122°F) each hour, until at room temperature (about 10 hours).

**Ideally, we would use a special kiln where the environment is finely controlled. However it can be managed by putting a lid on the crucible (less oxygen) – closer to a reduction atmosphere; or using the crucible without a lid (more oxygen) – oxidation atmosphere.

PURIFYING THE PIGMENT

1. Grind the sample using the pestle and mortar once more. Rinse with distilled water to remove uncombined sodium carbonate and sulfur, flushing several times through a coffee filter paper. Alternatively, a 10% solution of acetic acid and distilled water can be used to clean the pigment. Wash pigment thoroughly in distilled water afterwards until water runs neutral (check with pH indicator strip).

2. To remove the charcoal that was used as a catalyst in the process, flotation can be used. This is where an additive is added, such as a surfactant (see Natural Surfactants). A froth flotation process using natural olive soap can also be used to separate the component parts of the mixture, using approx. 1 part dry soap to 10 parts pigment.

DRYING THE PIGMENT

1. Once your pigment has been washed, filtered (and floated if doing), place your colours in a dehydrator at 70°C (158°F) for 4 hours. Spreading the wet pigment out thinly will speed this process up, but if allowed to dry in thicker clumps you may need to regrind the pigment using a pestle and mortar.

AN ARTIST'S ASIDE

THOUGHTS ON CREATING BLUE WITH MASTIC

Immense pressure and heat are needed over millions of years to form the metamorphic rock known as lapis lazuli. Synthetic ultramarine can be created by heating laboratory-grade raw materials in a kiln in a matter of hours. The temperature, proportion of ingredients and how much oxygen the mixture is subjected to throughout the process are a few of the variables that will create different shades of colour.

Thinking again about ultramarine's connection to mastic, which is used to purify the lapis stone to draw out the blue – this material has multiplicity in its connections to these two blue pigments. Mastic firstly reveals the blue, but then due to its use in varnishes and as an additive in painting mediums for glazing, it has the ability to shroud the blue over time as it reacts with oxygen and pollutants in the air – another symbolic chemical change in the context of paintings.

As an ingredient in chrism, used for spiritual libations, those tears of sap begin to take on a mystery that highlights our relationship with these materials. Libation, from the Latin *libare*, 'to pour as an offering', makes me think of those mastic drops, dripping from the tree.

PIGMENT SUMMARY

Pigment Name: Synthetic ultramarine

Nomenclature: Ultramarine blue, French ultramarine, Gmelin's blue

Pigment Index Number: PB 29

Colour: Intense blue, with a slightly violet undertone (ultramarine red shade)

Chemical Name: Sodium aluminium sulfosilicate

Chemical Formula: $Na_7Al_6Si_6O_{24}S_3$

Classification: Synthetic inorganic

Description: An affordable alternative to natural ultramarine (lapis lazuli), its vibrant blue colour is attributed to polysulfide ions trapped in an aluminosilicate matrix. Prized for colour intensity, lightfastness, and stability.

Conservation Issues: Sensitive to acids, which can degrade the pigment or cause discolouration. Prolonged exposure to acidic environments may result in colour shifts or fading.

Particle Morphology: Irregularly shaped, ranging in size from submicron to a few microns, contributing to its smooth texture and high coverage

Dates in Use: Introduced in 1826; remains in widespread use today

Density: Approx. 2.35–2.45g/cm³

Hardness: 5–5.5 on Mohs scale

Oil Absorption Ratio: 30–40g of oil per 100g of pigment

Refractive Index: 1.50

Health and Safety: Non-toxic pigment. Use appropriate safety gear when handling fine dusts.

Environmental Impact: Synthetic ultramarine has a lower environmental impact than natural lapis lazuli

NATURAL SURFACTANTS

Natural surfactants, derived from plants, animals, or micro-organisms, are surface-active agents that reduce surface tension between substances like oil and water, allowing them to mix effectively. When used with pigments, surfactants trap particles in foam, aiding in separation based on mass. This is particularly useful in levigation, enabling distinct pigment shades to be created from the same mineral source. Common natural surfactants include saponins, lecithin, and vegetable-based mild soaps, which are effective in purifying both synthetic and natural inorganic pigments.

COMMON SURFACTANTS

UK Sources: Lecithin from soya beans, sunflowers, and egg yolks.
Tropical Sources: Saponins from soapwort (*Saponaria officinalis*), soapberries, and yucca.
Synthetic Alternatives: Decyl glucoside or coco glucoside, made from corn and coconut.

CLEANING PIGMENTS

Surfactants remove impurities, grease, and residues through emulsification and dispersal without damaging pigment particles. For example, black and brown earth pigments often contain resinous impurities that can be washed away.

METHOD

1. Dissolve the chosen surfactant in water to create a 1% solution.

Note: Consider increasing the soap concentration if minimal foam is produced.

2. Mix the pigment gently with the solution using circular motions, avoiding airborne bubbles.

3. Allow the pigment to settle, pour off the solution, and rinse the pigment several times with clean water until soap residue is removed

FLOTATION AND SEPARATION

Foam generated by agitating a surfactant-pigment mixture can separate finer particles. The pigment adheres to bubbles and floats, while heavier particles settle.

METHOD

1. Use the same soap-pigment mixture from the cleaning process.

2. Agitate to generate bubbles; the surfactant stabilizes the foam, trapping finer particles.

3. Skim off floating pigment and rinse thoroughly to remove surfactant residue.

Note: The flotation process may vary depending on the sample. For example, in synthetic ultramarine production, fine carbon particles tend to be trapped in the foam, while heavier pigment settles.

DRYING THE PIGMENTS

After washing and separating pigments, place them in a dehydrator at 70°C (158°F) for 4 hours to dry completely.

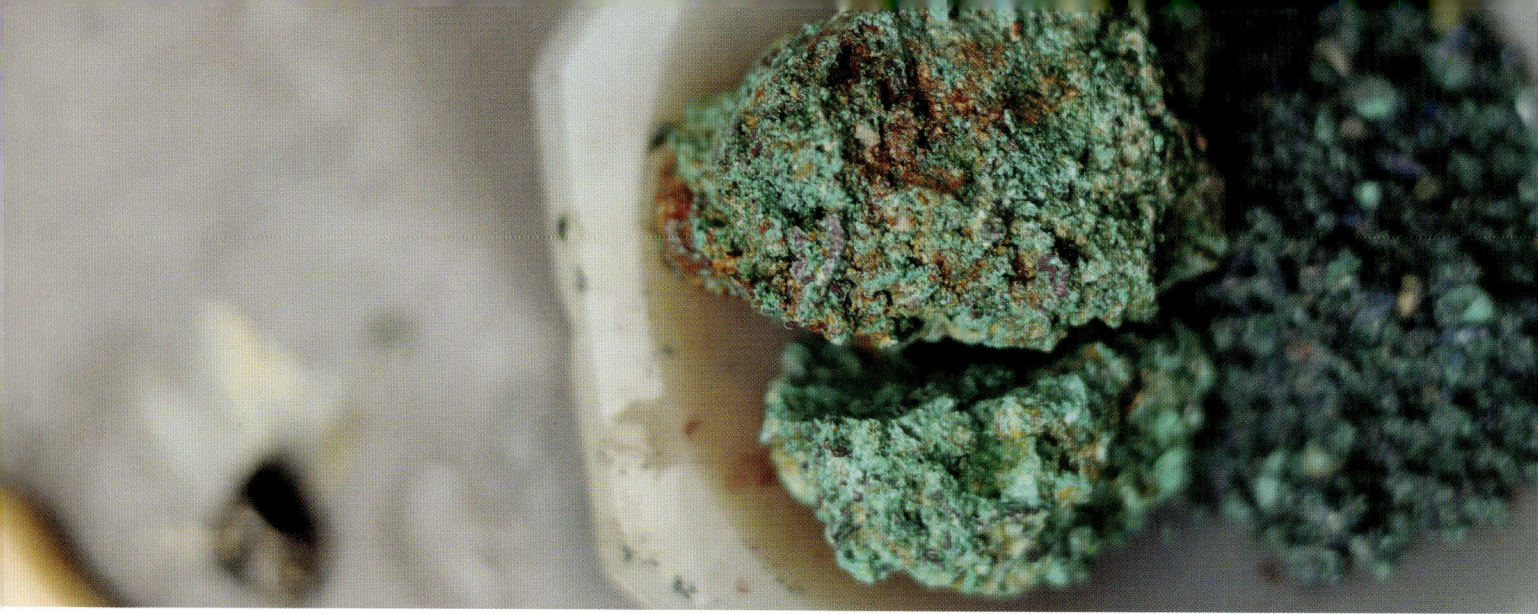

JOURNEYING WITH AZURITE AND MALACHITE

When Wignall & Moore, an architecture firm based in London, were working on a substantial refurbishment of the mineral gallery at the Royal Cornwall Museum in Truro, they contacted me to see if I could help. The gallery houses the museum's world-renowned collection of mineral specimens, with mahogany cabinets and specimens that were to be lit with specific directional lighting. For the walls, a dark blue-grey paint was desired. We spoke about mineral pigment paints, which led to a fascinating quest to produce a bespoke pigment mix using raw materials sourced from the local geology of the south-west of England to pay homage to the region's rich mining history and culture.

In a newly emptied, darkened mineral gallery, Jeni Woolcock, Collections and Engagement Manager, showed me 40 or so boxes of potential minerals. Amongst many unusable specimens (some were toxic, or would not produce a colour when ground, or were too hard), there were about 7 kilos of material that contained azurite (a soft, deep-blue copper mineral) and malachite (a bright-green copper carbonate hydroxide mineral). These azurite and malachite pigments were created from rocks (above) from County Galway, Ireland – waste materials from copper ore mining from a mine of unknown provenance – and were part of the museum's research collection. I had the starting point for my experiments to create a bespoke pigment blend for the wall paint.

Jeni also shared her contacts with local mines and quarries in Cornwall, and I visited a number of these sites. On visiting Cornish Lithium in St Dennis, I realized that kaolin (hydrated aluminium silicate) was not only a useful filler in paints but was also an integral ingredient in synthetic ultramarine manufacture. I was shown two large waste heaps of kaolin, affectionately named by local residents as 'flatty' and 'pointy'. The kaolin here was a relatively pure mineral – a bright white clay, sometimes with streaks of iron staining. Granite, quartz, and feldspar were being removed as part of prospecting for high-grade lithium for use in battery cells. As a by-product, the kaolin was a useful waste stream material to use in my experiments.

Imerys Minerals, in the village of Par, sent me the purest grades of kaolin that they routinely sell to commercial pigment-making companies. I experimented with their kaolin to make different shades of ultramarine blue and green. One site

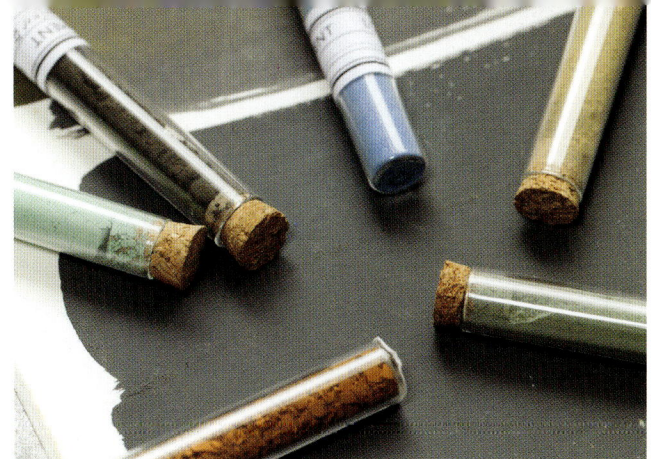

I couldn't go to, due to time constraints, was the Delabole Slate Quarry. However, they generously supported the project with several kilos of slate powder in a soft, silvery grey-blue colour, as well as raw slate chips for the museum's pigment display. The last site visit was the South Crofty tin mine project (near Camborne) operated by Cornish Metals; the radioactive levels emanating from the mine that day were not safe, so we visited their new site instead. Here a series of gigantic tanks take the pumped-out mine water through a labyrinth of piping. Acids and alkalis are added to remove arsenic and other heavy metals picked up by the water in the mine, which can cause degradation to aquatic life and have negative impacts on local ecosystems, and the water is made safe and neutral. A by-product of this process is iron oxide, a useful and stable mineral pigment.

I experimented for days with the proportions of the pigments made from the Cornish sites, and eventually came up with a dark grey-blue-green colour, very close to what the architects originally wanted. I started with the slate powder as the base, adding synthetic ultramarine blue to create a soft blue-grey. I added carbon black, a nod to its inclusion in the synthetic ultramarine recipe, to darken the tone of the colour and knock back the vividness of the blue. Lastly, a little waste stream iron oxide pigment and kaolin were added for an 'earthier' feel, creating the colour you see in the colour swatch. The paint binder was a breathable emulsion made from casein, a milk derivative, made for us by Rose of Jericho – a leading manufacturer of traditional paints.

A permanent display of the craft of pigment making was created at the museum to tell the story of how the paint was researched and formulated, including raw azurite and malachite stones alongside their processed pigments. The museum's selection of mineral specimens available for pigment processing include those that have degraded over time, or are no longer useful for the collection to keep intact. Most were bequeathed by Lesley Nash, a late 20th-century Cornish dealer and mineral collector. At one point they were being sold in the museum shop, but no one wanted them. I'm glad they are being put to use.

AN ARTIST'S ASIDE

MINERAL PIGMENT TECHNIQUES

Rock that bears azurite, malachite, and other impurities, such as quartz, is pulverized using a pestle and mortar, with distilled water to dampen it to stop airborne dusts. As iron oxide and malachite have a lower density to azurite and are softer, they are suspended in the water; they can be separated by allowing the azurite to sink to the bottom of the vessel, then pouring off the water laden with other minerals. This process is known as levigation (see Chapter 6). I think about the particles levitating in the water, hovering above the materials they were just attached to. They had been attached to each other for a long time, maybe thousands if not millions of years.

Reflecting on this process, I wonder about the ethics of how, once crushed, these materials cannot be put back together. To destroy these materials and separate them into distinct parts serves to produce non-toxic colour and a meditative craft practice, but who am I to fast forward time, weathering these rocks with my hands and tools? This emulates long geological processes, speeding them up to happen in a matter of moments. The pigment maker's studio exists, then, as a kind of time vacuum. The destructive and extractive nature of creating mineral pigments makes me pause and think carefully about each material and project I encounter, not only in terms of the processing but also the sourcing of the raw material.

RECIPE: AZURITE

The processing of azurite and malachite pigments primarily involves separating the minerals from their host rock and impurities through careful grinding and washing. Depending on the purity of the raw specimen, up to 50 washes may be needed to fully refine the pigments. It is an incredibly satisfying task, as with each successive wash, the blue or green hues emerge more vividly from the muddy waters.

INGREDIENTS

- Azurite*
- Distilled water
- Acetic acid or white vinegar
- Rabbit skin glue granules, approx. 30g per 300ml water

EQUIPMENT

- PPE – respirator (P3 rating), gloves and safety goggles
- Hessian sack and reinforced plastic bags
- Pick and metal mallet or hammer
- Magnifying glass
- Inert spatula (plastic, stainless steel, glass)
- Pestle and mortar (laboratory-grade porcelain to prevent staining)
- Glass petri dishes for sampling
- Coffee filter papers
- pH indicator strips
- Large bowls or 1 litre paint kettles
- Geological sieves in various grades (40, 60, 80 and 100 mesh)
- Dehydrator or domestic oven

Source from a reputable mineral dealer, who has proof of ethical standards and provenance.

METHOD

GRADING THE MINERAL SAMPLE

1. Inspect the specimen. If the sample contains finite areas of different colours and minerals, break it apart carefully using a hammer and pick. Once these initial samples have been broken up, they can be sorted into rough colour groupings.

2. Place like samples in a thick plastic bag (two layers would work well) and then into a hessian sack. Smash with a hammer or metal mallet until all pieces are roughly pea-sized.

3. Using a magnifying glass, sort the pieces again into colour groups, as follows:

- Blue: coarse azurite
- Green: coarse malachite
- White: calcite, limestone, quartz, chalk
- Orange: iron oxide
- Black: manganese oxide

Put aside any colours you are not interested in. These can be returned to the earth. (Note: if your sample contains more malachite, the method described can be employed to purify this as the main colourant extracted.)

REMOVING IRON IMPURITIES

1. Mix the coarse azurite lumps with distilled water using a pestle and mortar and agitate, to help to rinse off iron oxide and other impurities that can become attached to them. By repeatedly rubbing in a circular motion with a gloved finger, the azurite can be cleaned without breaking down the particle size, which would desaturate the colour. Pour off the muddy brown water to reveal the blue azurite settled below.

2. Repeat this cleaning process until the water is relieved of its brown or orange tinge. It should turn increasingly blue-green.

REMOVING FURTHER MINERAL IMPURITIES

1. Break down the lumps of cleaned blue azurite and grind in more distilled water using a pestle and mortar, and continue to do this until the colour of the water changes. It will most likely go from muddy brown, to deeper orange and then to blue-green. Repeat until you have several 'grades' of pigment.

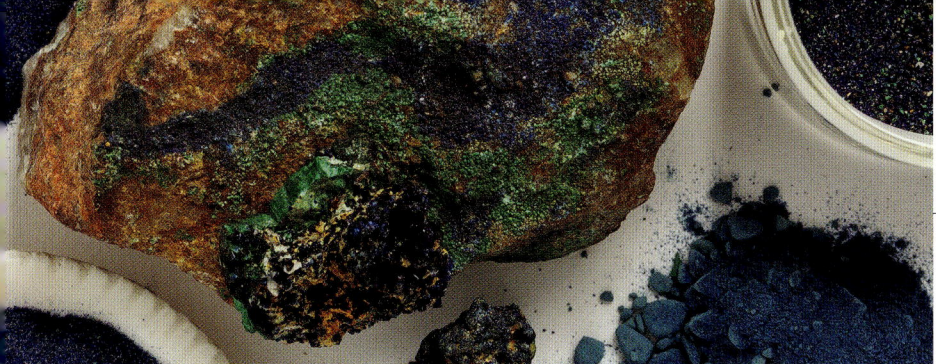

2. The addition of a liquid animal glue to the distilled water will aid the separation of the particles. Add a solution (10 parts water to 1 part glue) to the coarse azurite sand. Grind until the solution becomes murky with suspended fine particles, then allow it to settle and the glue to coagulate until the following day.

3. The next day scoop out the jellied material and manually pull the layers apart. These layers can then be mixed with warm water and further washed to remove the glue solution. The process of flotation can be used for this (see Natural Surfactants), or indeed as an alternative to using rabbit skin glue at all.

DRYING AND GRINDING

1. Now you will be left with a coarse blue azurite grit. Grind this until you have reduced it to a 40-mesh powder. This can be checked by putting it through a 40-mesh sieve.

2. Wash the powder several times: the water should be clear now with no bubbles or foam (these indicate impurities are still present).

3. Continue to grind samples of your azurite pigment to 60, 80, and 100 mesh powder samples – the colour will desaturate to a lighter blue with a green leaning. During this process, the water may also turn more of a green-blue, which will be a mix of malachite and azurite, a very beautiful colour that can be poured off and separated.

4. Place the azurite pigment samples in a dehydrator to dry at 70°C (158°F) for 6 hours, or dry in the oven on the lowest setting. Alternatively, spread the pigments thinly on an impermeable surface and let dry in the sun or on a radiator. Once dry, bottle and store for later use.

TIP

Geologists clean azurite specimens with acid solutions and distilled water, removing malachite in situ to form green copper solutions that rinse away easily. Samples can be cleaned with white vinegar or acetic acid, and this step can be utilized after washing as much iron oxide impurities away with water. This milder method requires thorough rinsing in clean water to create a neutral, stable material.

PIGMENT SUMMARY

Pigment Name: Azurite

Nomenclature: Copper(II) carbonate, mountain blue, Azurro della Magna

Pigment Index Number: PB 30

Colours: Blue, deep blue, light blue leaning to green hue

Chemical Name: Copper(II) carbonate

Chemical Formula: $Cu_3(CO_3)_2(OH)_2$

Classification: Natural inorganic

Description: Derived from copper ore; valued for its vibrant blue hue

Conservation Issues: Highly sensitive; exposure to light and air can cause it to alter chemically into malachite, a greenish pigment. Prone to fading and chalking; may become brittle, especially in humid or acidic environments. Strict light and humidity controls required.

Particle Morphology: Crystalline, granular, with a fine texture

Dates in Use: From antiquity, with peak use from the Renaissance to the 19th century

Density: 3.8–4.3g/cm³

Hardness: 3.5–4 on Mohs scale

Oil Absorption Ratio: 20–25g of oil per 100g of pigment

Refractive Index: 1.69–1.70

Health and Safety: Non-toxic in its stable form but can be harmful if inhaled as a dust

Environmental Impact: Extraction of azurite from mines, particularly in regions like Mexico and Morocco, has significant impact

JOURNEYING WITH VERDITER PIGMENTS

Blue and green verditer are artificial pigments used widely in European painting and decorative arts during the 17th and 18th centuries – particularly for underpainting and shading, and in areas where brighter hues were needed but more expensive pigments such as ultramarine or malachite were prohibitive. They were likely to have been discovered in Europe during the late Renaissance. The term 'verditer' originates from the French *vert-de-terre*, meaning 'green of the earth', though it eventually referred to artificially created pigments.

These pigments emerged as by-products of the silver-refining process. Copper was often dissolved in nitric acid during silver refining, and when lime (calcium hydroxide) was added to the resulting solution, copper carbonate precipitated out. This precipitate, when carefully processed and washed, became blue or green verditer.

In addition to their use in art, verditer pigments were used in the industrial production of inks, ceramics, and other decorative finishes, thanks to their bright, opaque colours. Their chemical compositions as basic copper carbonates made them relatively stable and vibrant, offering affordable and consistent alternatives to naturally occurring mineral pigments. By following recipes that precipitated copper carbonate from solutions of copper sulfate or nitrate, manufacturers could produce these pigments in large quantities, contributing to their widespread use in art and decoration.

After visiting Van Eycks' *Adoration of the Mystic Lamb* altarpiece (completed in 1432) in Ghent, Belgium, I was intrigued by the green pigments used by the Van Eyck brothers to create the vibrant, luscious colour of the grass and foliage. In my research I came across discussions in the *Pigment Compendium* (see Bibliography) about the production of green and blue verditer pigments, which raised the possibility that these colours might have been present in the *Mystic Lamb* triptych. This discovery surprised me, as it predated what I had previously understood about the production of blue verditer, that I believed emerged in the early 1700s.

This highlights how complex researching historical pigments can be – many pigments have histories stretching far beyond what is commonly documented. Most of the writing available on pigment manufacture focuses on later, large-scale commercial efforts, yet pigments like these were probably produced on much smaller, personal scales by artists and alchemists centuries earlier. It's this intimate, experimental side of historical pigment creation that I find especially compelling, as it ties modern experimentation with colours directly to the innovative practices of those earlier artists.

Green verditer first appeared in Western oil-painting techniques during the 1500s. It shares an identical chemistry with genuine malachite pigment, which is derived from the natural stone, but it lacks the mineral impurities found in malachite, and structurally their crystals are of a different shape and size. With a tendency towards a spherical structure, green verditer is more uniform than its naturally occurring counterpart, also paler and often less opaque.

RECIPE: 18TH-CENTURY ENGLISH BLUE VERDITER

My early experiments with verditer pigments included using calcium carbonate (chalk) and calcium hydroxide, which created a softer, less stable blue. This recipe was devised from an amalgamation of recipes including those by Keith Edwards and The Alchemical Arts. The blue is accentuated with each wash, and you can also experiment with temperature and timings to create different shades.

INGREDIENTS

- Calcium carbonate, 6g and 4.5g
- Copper nitrate*, 15g and 11.25g
- Distilled water

EQUIPMENT

- PPE – respirator (P3 rating), gloves, and safety goggles
- 500ml glass beakers
- Inert stirrers (glass, plastic, stainless steel)
- Coffee filter paper and plastic funnel
- Dehydrator

Copper nitrate can be substituted with less toxic copper sulfate, but it creates a more unstable pigment.

METHOD

PRECIPITATING THE PIGMENT

1. Dissolve 15g of copper nitrate in 200ml in cold water to create a transparent blue solution.

2. Slowly add 6g of the calcium carbonate to the solution while stirring – be careful not to splash. The solution will be an opaque light blue colour

3. The reaction will form a blue precipitate, which will start to settle at the bottom of the container – this is copper carbonate hydroxide. For 12 hours, every 30 minutes stir for 2 minutes.

CHEMICAL SAFETY PRECAUTIONS

Store in a clearly labelled airtight inert container, away from children and pets.

On contact with skin, wash immediately with water.

On contact with eyes or if accidentally ingested, seek immediate medical attention.

If dust is inhaled, seek immediate medical attention.

PIGMENT SUMMARY

Pigment Name: Blue verditer

Nomenclature: Blue bice, cuprammonium blue, refiner's verditer, mountain blue, ashes blue, cendres bleue, cendres bleu d'angleterre, air blue, lime blue, neuwied blue, copper oxide

Pigment Index Number: PB 30

Colours: Light to dark blue with green leaning, semi-opaque

Chemical Name: Basic copper carbonate

Chemical Formula: $Cu_3(CO_3)_2(OH)_2$

Classification: Synthetic inorganic

Description: A synthetic form of azurite, consisting of a copper carbonate hydroxide

Conservation Issues: Can darken or change colour when exposed to acids or sulfide gases. Most suitable for water-based applications (tempera, watercolour, gouache); in oil, will eventually turn green.

Particle Morphology: Large, crystalline structure

Dates in Use: Medieval period to 19th century

Density: 3.7g/cm³

Hardness: 3.5–4 on Mohs scale

Oil Absorption Ratio: 23g of oil per 100g of pigment

Refractive Index: 1.70

Health and Safety: Copper carbonate is classified as hazardous, with mild skin and eye irritability

Environmental Impact: Significant as made from copper compounds, which may involve mining of metals. Dispose of waste materials responsibly to prevent harm to aquatic environments.

4. Leave to sit overnight and repeat step 3 for the next two days. Subtle bubbling or frothing will occur. The solution can be kept cool in the fridge. Over time the solution will become a lighter blue and the pigment will become darker.

5. Pour the excess liquid into a separate container and add water to dilute it before pouring down the drain. Scoop the remaining pigment into a coffee filter paper placed in a funnel over a beaker. Wash the precipitate thoroughly with distilled water to remove impurities.

6. Dissolve 11.25g copper nitrate in 150ml water, then add the freshly made pigment with 4.5g of chalk. Repeat steps 3–5. (If you have time, this can be repeated to make a more intense blue colour.)

DRYING AND GRINDING

1. Dry the blue powder in a dehydrator or by spreading the wet pigment on an impermeable surface. Once dried, grind it to the desired fineness.

RECIPE: 18TH-CENTURY EUROPEAN GREEN VERDITER

In my experience, green verditer is much easier to make than blue. Additionally, blue verditer, if not thoroughly washed during the production process, can lighten and eventually turn green over time. Various 18th-century recipes describe reacting copper solutions with different reagents – most commonly sodium carbonate, sodium bicarbonate, or potassium carbonate (potash) – to produce these pigments. This recipe is consistent with descriptions found in early 18th-century European treatises on pigment making, such as *The Painter's Companion* (London, 1740), whose author is unknown.

INGREDIENTS

- Copper sulfate, 20g
- Distilled water
- Sodium bicarbonate, 20g

Alternatively, use 15g of chalk.

EQUIPMENT

- PPE – respirator (P3 rating), gloves, and safety goggles
- 500ml glass beakers
- Inert stirrers (glass, plastic, stainless steel)
- Coffee filter paper and plastic funnel
- Pestle and mortar
- Dehydrator

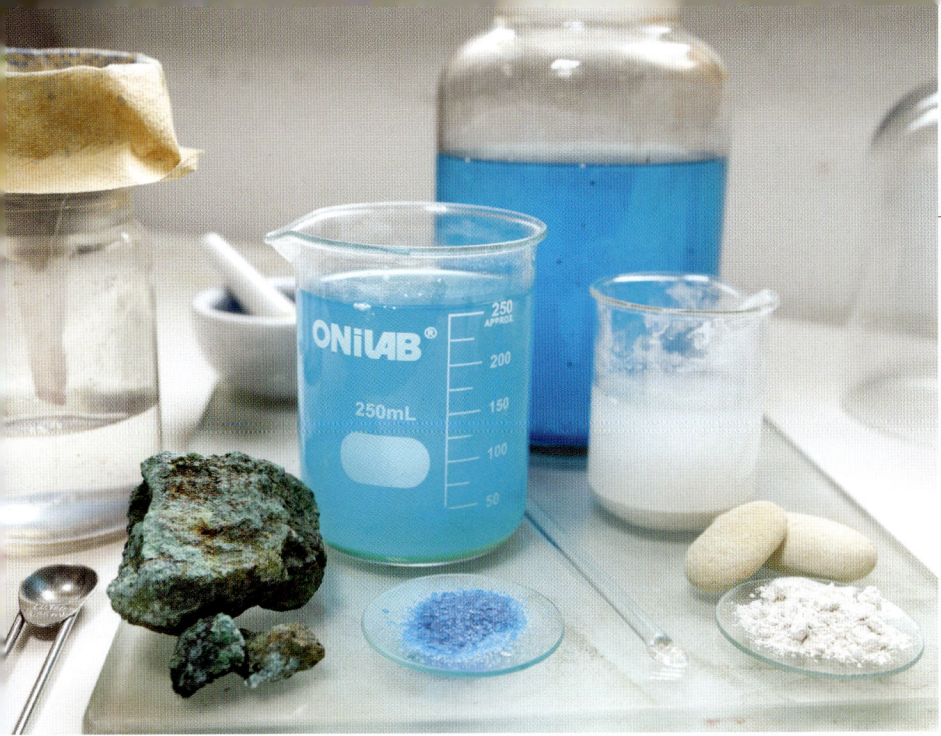

METHOD

PRECIPITATING THE PIGMENT

1. Dissolve 20g of copper sulfate in 200ml of just boiled (80°C/174°F) distilled water. Stir well to get a transparent blue solution.

2. Dissolve 20g of sodium bicarbonate into 200ml of hot distilled water.

3. Slowly add the bicarbonate solution to the copper sulfate solution whilst heated on a hob at around (80°C/174°F). Do this slowly as lots of foaming will occur. Stir continuously as they are combined. (Ideally, use a magnetic stirrer on an electric, temperature-controlled hotplate for best results.) Keep stirring until no more frothing is created, then add some fresh hot distilled water (50ml) and stir again for 3 minutes.

4. The precipitated pigment is now formed and a bright green will be produced.

5. Leave the solution to cool and the pigment will settle out completely.

FILTERING THE PIGMENT

1. Filter the pigment using a coffee filter paper inside a funnel over a beaker and flood the pigment with several washes of hot distilled water to remove any unreacted materials.

DRYING AND GRINDING

1. Dry the green powder in a dehydrator or by spreading the wet pigment on an impermeable surface. Once dried, grind it to the desired fineness.

PIGMENT SUMMARY

Pigment Name: Green verditer

Nomenclature: Synthetic malachite, green bice

Pigment Index Number: PG 39

Colour: Bright, semi-opaque green

Chemical Name: Basic copper carbonate

Chemical Formula: $Cu_2(CO_3)(OH)_2$

Classification: Synthetic inorganic

Description: Green verditer is the historical name given to synthetic green copper carbonate hydroxide with a chemical structure identical to the naturally occurring mineral malachite. Unlike the natural mineral, it adopts a distinctive spherical crystal structure (spherulitic habit).

Conservation Issues: Can darken or change colour when exposed to acids or sulfide gases. Will decompose in hot water, above 200°C (392°F).

Particle Morphology: Large, crystalline structure

Dates in Use: 15th to 19th century

Density: 4g/cm³

Hardness: 3.5 –4.0

Oil Absorption Ratio: 25g of oil per 100g of pigment

Refractive Index: 1.655

Health and Safety: Can be toxic, especially if inhaled or ingested in large quantities; handle with care

Environmental Impact: Shares similar environmental concerns to blue verditer. Care should be taken to prevent contamination of ecosystems.

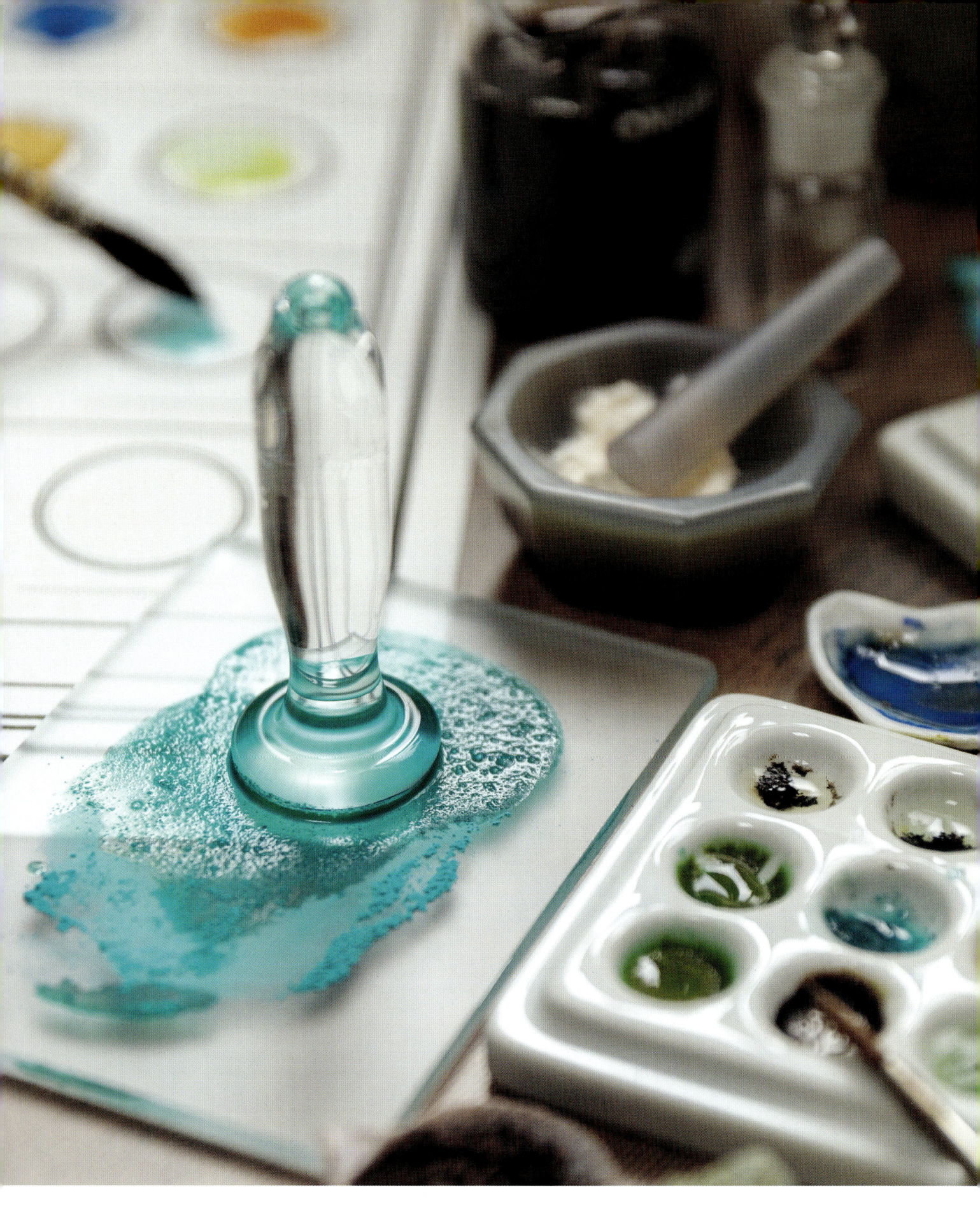

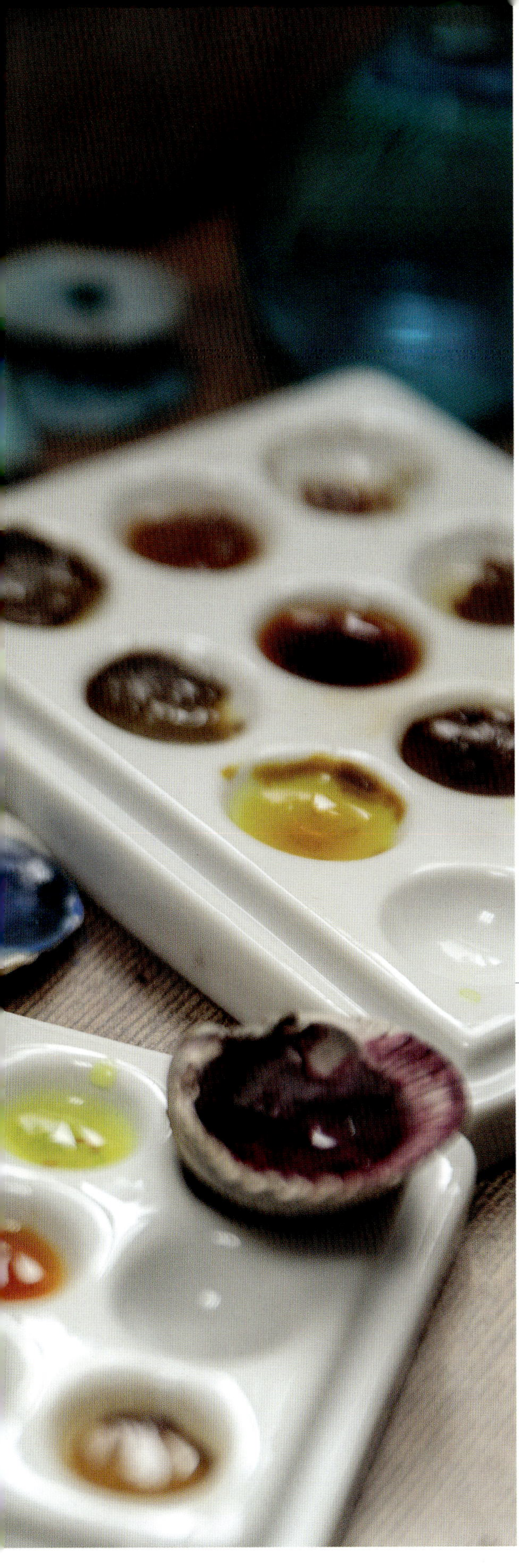

CHAPTER FIVE

THE MIXING OF PIGMENT

The art of mixing pigments is a fine skill passed down through the centuries. Whether grinding minerals by hand or blending synthetic colours in a lab, the process has been essential for achieving the perfect hue, tone, and texture as demonstrated by the historical palettes laid out presently. This chapter focuses on the techniques and tools used to prepare and mix pigments and paints, from traditional methods involving pestle and mortar to modern methods employed by artists and manufacturers today. Additionally, the chapter delves into how pigments interact with different binders and media, exploring how mixtures are created to achieve the desired visual and physical properties in paints and other art materials.

CREATING HAND-PAINTED PIGMENT SAMPLE COLOUR CHARTS

Colour charts are invaluable tools for organizing, comparing, and contrasting your handmade pigment recipes, while also tracking their colour changes in different binding mediums. To monitor how ultraviolet light affects your colours, consider attaching a small paper flap over half of each colour sample. This simple method allows you to observe and document any fading or shifts over time.

Colour swatches for these colour charts are painted in casein paint, at full opacity, for the densest mass tone. Casein paint is created by suspending pigment in a protein-based binder known as casein, which is derived from the solid whey found in milk.

Full opacity displays the structure of each individual particle, as there is no excess binder to obscure the quality of each pigment. The structure relates to individual dimensions of the grain in terms of size and

form, which affect how light is refracted throughout the paint layer. These variables can change the colour greatly, from the opacity and tinting strength to its saturation.

Unlike Pantone colours, for example, hand-created pigments allow you to experience the complex chemistry and ephemeral nature of the original plant, for example, such as seasonal colours and other variations. Casein binder, distinct from other binders such as oil, does not obscure the unique conglomerate qualities of individual particle structures, so it could be the one to mix your pigment. However, hard, insoluble casein contains ammonium carbonate, which is alkaline and can cause some pigments, such as Prussian blue, to be bleached out. There is no one binder that is suitable for all pigments. Utilize the casein recipe (see Recipe: Casein) to create your own colour chart or palette.

Below: Testing the hue and saturation of two coal tar derived pigments (azo crimson and dioxazine purple) by mixing them with chalk and carbon black.

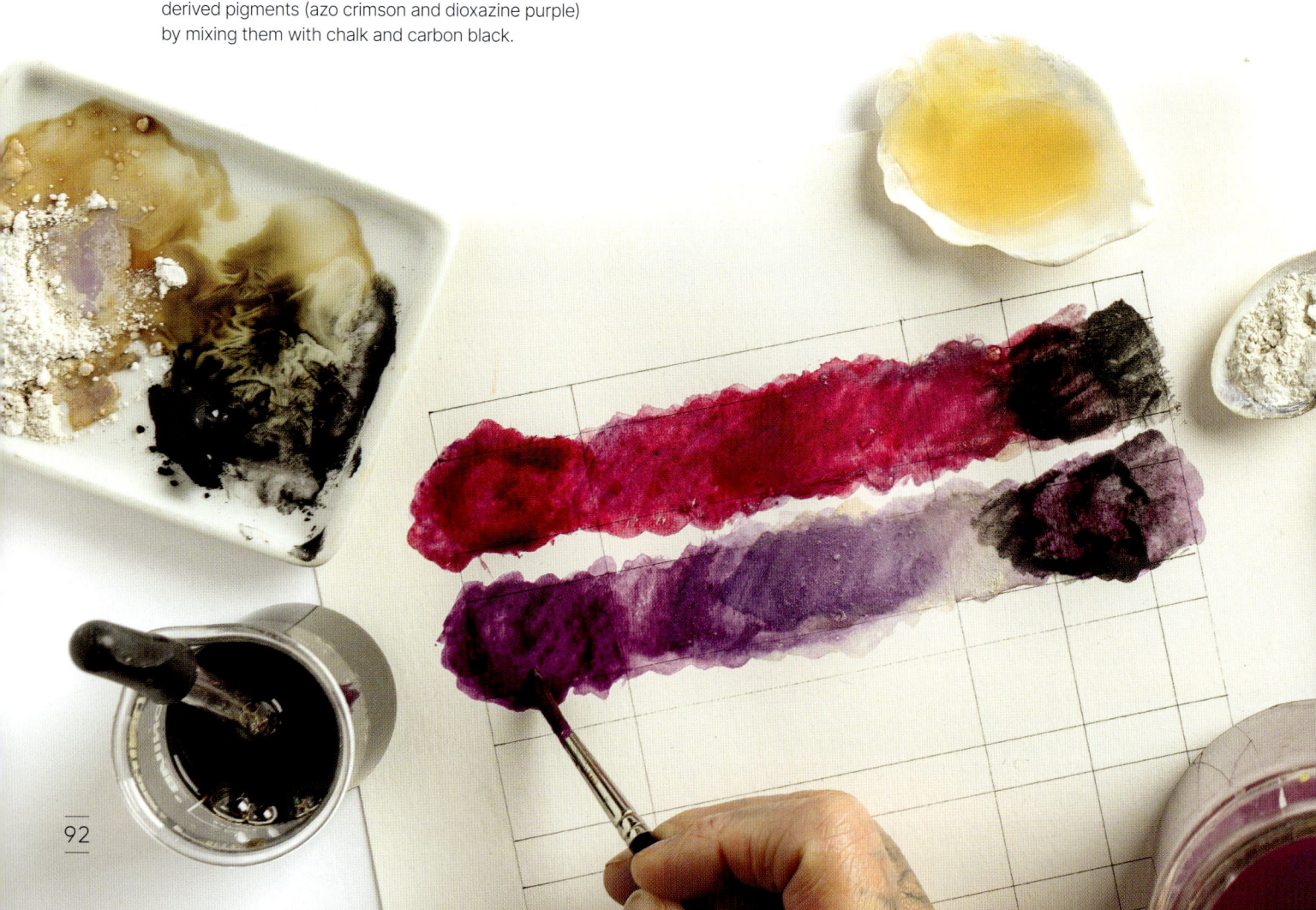

AN ARTIST'S ASIDE

THOUGHTS ON PAINTINGS AND PIGMENTS

Paintings are the accumulation of materials that reflect the human condition and are space for complex ideas to play and unfold. The patina of life: stains, dirt, residue snooker chalk, pigments are in service of mediums and art forms. Pigments can be viewed as vehicles for ideas, carriers of information, of space, process, human touch suspended in glue. They contribute to creating visible worlds that invite immersion through colour, yet are composed of the world itself. An object might be portrayed with a piece of itself ground into powder, to be considered a complete representation of something. For instance, a painting of a safflower plant could be created using safflower pigment suspended in safflower oil, painted on paper made from pulped safflower – a portrait of the material in its entirety.

A painted surface is a culmination of substances found from all over the world. The sourcing of pigments reflects the changing times, the contemporaneous fluctuations in politics and economics, the geographical and social changes that affect price and availability. They can be challenging to locate, such as lapis lazuli from war-torn Afghanistan, or manganese blue from drying-up mines in the US and China.

Paint is made from binders mulled with powdered particulates: man-made, naturally sourced, organic, inorganic, processed chemically, synthesized in the laboratory. A painter's palette, with all those colours, is perhaps rather more than just paint. The word 'paint' is a false homogenizing nomenclature; the overlooking of the unique sources of our colour denudes the material's unique expression.

The aesthetic reception of colour – the sensations we derive or receive from a surface that has been modified with a colourant – are not abstracted moments. They are inextricably linked to their provenance. Paintings are organized, stratified vestibules of matter. They exist as surfaces rendered by the capturing of literal matter. The hierarchies of materials from specific time periods are made manifest and are revealed by their incorporation into the picture's surface. These are a reflection of the tastes and choices of the painter, but also to the availability of the material. These organized particles are coaxed into forms that communicate vast arrays of emotions. A spell is cast using ingredients from far-flung places, centripetally converging in one place. The artist is a conjurer. Human labour and time is concentrated with intention, intensifiying the object and giving it a special resonance.

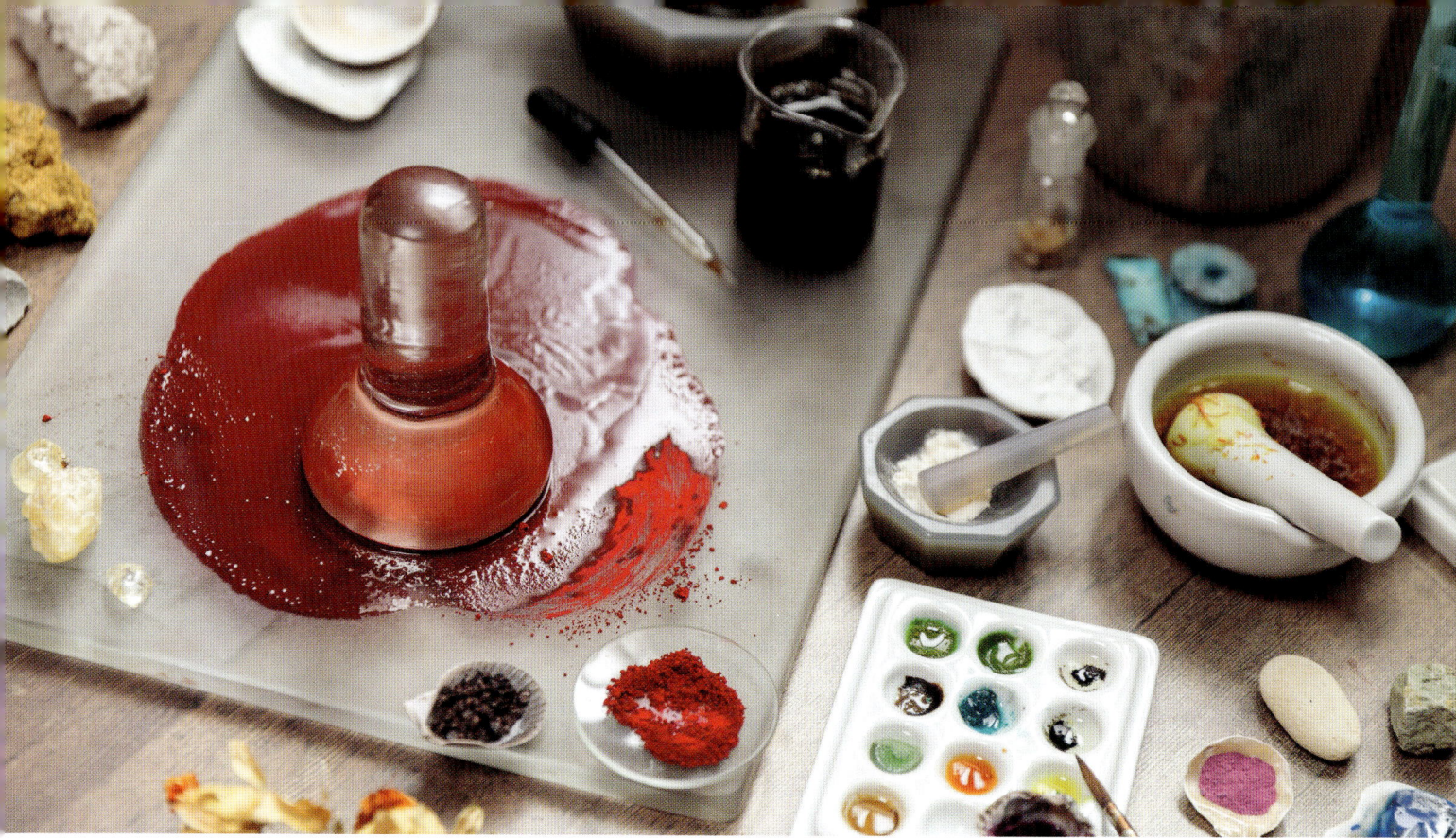

Pigment making is an activity that takes you out of the commercial sphere. It replaces a distant and de-sanitized way of working: a handmade pigment is less altered from its natural or original state in order to access the material's true or unmodified properties. As such, it may prompt deeper thinking about societal patterns of commerce.

The reduction of a specific event with all its details – noises, textures, differences in material, cause and effect – into a collection of colours is the process of extracting one form of reality (the original materials) into a uniform matter. Abstracted, in glass jars, pigments are meaningless apart from the optical qualities that they evoke. Yet in reducing something to its purest form, a new kind of documentation arises.

Pigments used by artists throughout history have been made from such disparate materials as flowers, rocks, and beetles. Yet paint manufacturers have tried hard for hundreds of years to disconnect the artist from the truth of their materials. Colours with specific provenance have been homogenized into paint ranges that use fillers and additives to mask the unique handling qualities of the pigments.

A thrill in hand-making pigments is that of the infinite possibilities of a material. Anything can be repurposed, recycled, and collaged. Pigment making is an activity where the end goal is to make a fine-grained particulate with some colouring power – but this is almost an excuse for the labour of the process, for the explorative and rhythmic technique used in order to coax a colour into being. Naming plants and placing them within categories creates an understanding, which leads to a better sense of purpose. Rather than walking blindly down the street, we can see what was there already in front of us. It's about taking value from the mundane.

Observing phenomena in the studio reveals phenomena in the outside world, and vice versa. For example, particles in rivers and seas are emulated in the rise and fall of pigment particles in the pestle and mortar. Everything – every surface, every material – is in a state of flux, and so it's the visible and tangible changes that alert us to the make-up of materials in the studio.

Above: Carmine pigment being mulled into a watercolour.

RECIPE: CASEIN

Casein is a protein found in milk, making up about 80% of its total protein content. It creates a hard-wearing paint that can be used for works of art and for interior finishes on walls.

INGREDIENTS

- Skimmed milk, 1 litre
- White vinegar, 2–4 tablespoons
- Ammonium carbonate, 1.5g* (5% of the casein powder weight)
- Distilled water
- Dry pigment

EQUIPMENT

- 1.5 litre cooking pot
- Stirring spoon
- Thermometer
- Measuring spoons
- Muslin cloth and kitchen sieve
- Mixing bowl
- Clean cloth or paper towels
- Dehydrator (optional)
- Storage jar

Alternatively, use 5g of potassium carbonate, and dissolve in 100ml warm distilled water.

METHOD

MAKING CASEIN POWDER

1. Pour the milk into the pot and gently heat to around 50–60°C (122–140°F). Do not boil.

2. Slowly add the vinegar while stirring. The milk will begin to curdle as the casein separates from the whey.

3. Once the curdling is complete, pour the mixture through a muslin cloth placed in a kitchen sieve to separate the solid casein from the liquid whey.

4. Rinse the curds with cold water to remove excess acid, then press out extra moisture using a clean cloth.

5. Spread the curds on a flat surface (such as a tray) and allow to dry in a warm room (or use a dehydrator). Store the dried casein powder in a jar.

MAKING CASEIN BINDING MEDIUM

1. Weigh 30g of dry casein powder and place in a small glass or ceramic container.

2. Measure 1.5g of ammonium carbonate (i.e., 5% of the casein powder weight).

3. Dissolve ammonium carbonate in 30ml warm distilled water. Stir gently.

4. In a well-ventilated space, slowly add the dissolved ammonium carbonate to the casein while stirring continuously. The solution will start to foam slightly as ammonia gas is released.

5. Continue stirring until the casein fully dissolves into a smooth liquid binder. If needed, adjust the water amount to create the desired consistency of a light gel.

6. Let it rest for a few minutes before using to allow any remaining gas to escape.

MAKING CASEIN PAINT

1. Using a pestle and mortar combine your pigment with the casein gel in a 1:1 ratio to form a thick paint, adding water to achieve the desired consistency of single cream. Larger quantities can be ground into the medium using a muller and glass slab.

PREHISTORIC COLOURS OF EUROPE

The palette used by the earliest humans in the UK and Europe was primarily composed of natural earth pigments such as red and yellow ochre, charcoal, white chalk, and black manganese dioxide. These pigments were readily available in the natural environment, and they were durable and versatile. Their use in cave paintings and body adornments highlights the early humans' understanding of their surroundings and their ability to express themselves through art.

In Europe, the earliest use of pigments dates from the Upper Palaeolithic period, around 40,000 years ago, to the Mesolithic period (up to around 8,000 BCE). The Clearwell Caves in Gloucestershire contain the longest-used ochre mine and processing site in Europe, with evidence of ochre extraction dating back 4,500 years. The methods developed for grinding, mixing, and applying pigments in these early times continue to astonish archaeologists today.

The Upper Palaeolithic period marks the appearance of the first anatomically modern humans (*Homo sapiens*) in Europe and the beginning of sophisticated artistic expression. The earliest evidence of pigment use in Europe comes from Upper Palaeolithic cave paintings. The Chauvet Cave in France, which dates to around 36,000 BCE, is one of the oldest known examples of figurative art, showcasing red ochre, yellow ochre, black charcoal, and manganese dioxide. Upper Palaeolithic humans also used pigments for body adornment and to decorate portable objects, such as carvings and figurines.

The Mesolithic period follows the end of the last Ice Age, when hunter-gatherer societies continued to use pigments for decoration and symbolic purposes. The use of ochres and other natural pigments continued into the Mesolithic period, as evidenced by archaeological finds of ochre-stained tools, burial sites, and rock art in various locations across Europe.

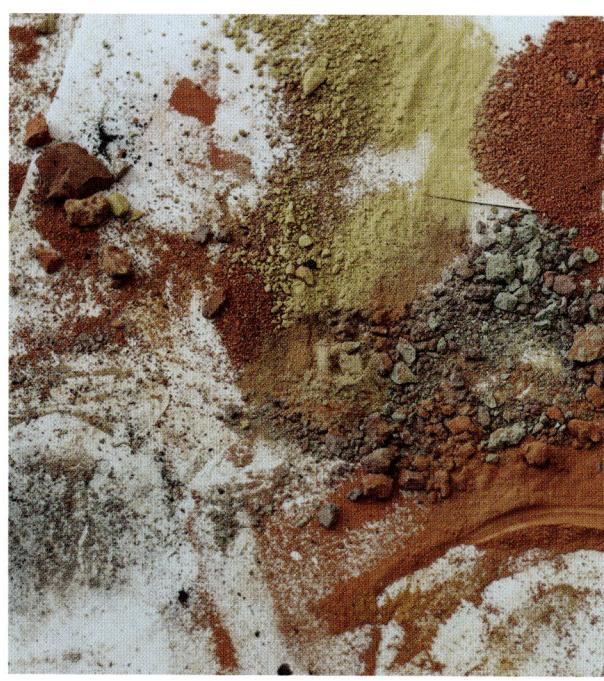
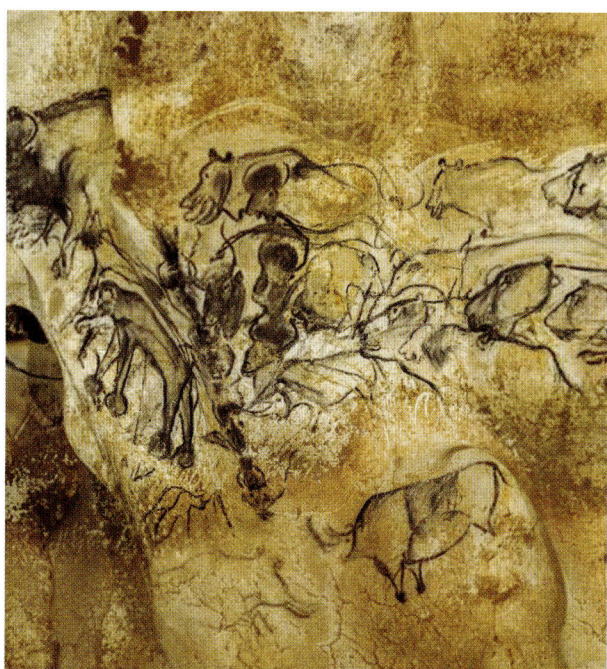

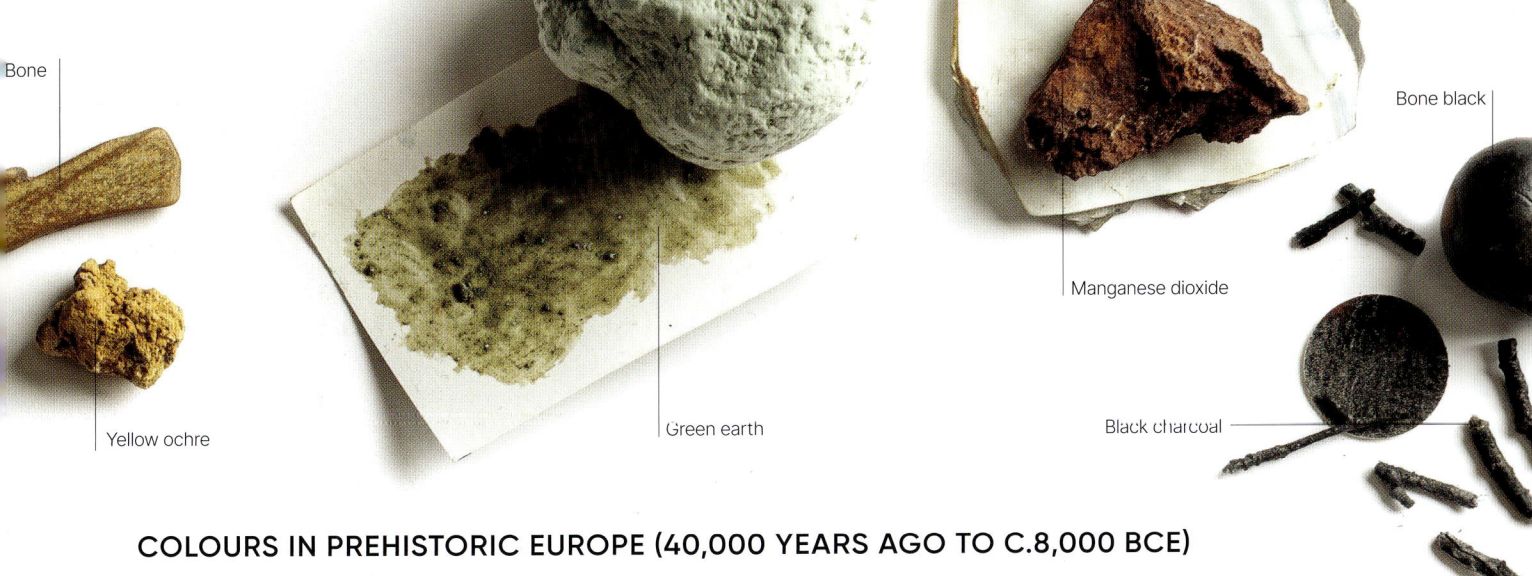

Bone

Yellow ochre

Green earth

Manganese dioxide

Bone black

Black charcoal

COLOURS IN PREHISTORIC EUROPE (40,000 YEARS AGO TO C.8,000 BCE)

Red Ochre: An iron oxide (hematite) and one of the most widely used pigments. Found throughout Europe, particularly in areas with iron-rich soils, it was used for cave paintings (Chauvet Cave, France), burial rites (ochre was sometimes sprinkled over the dead), and body decoration. *Colour:* Wide range of red hues, from deep red to orange, depending on its mineral content and how finely it was ground.

Yellow Ochre: Another iron oxide (goethite) abundant in many parts of Europe, particularly in areas with rich clay deposits. It was used similarly to red ochre in both cave paintings (Lascaux Caves, France) and as a body paint. *Colour:* From pale yellow to brownish-yellow, depending on the conditions in which it formed.

White Chalk: White pigments were often derived from chalk or kaolin clay, naturally occurring in limestone regions across Europe. Examples of white chalk can be found in cave paintings such as those in France's Cosquer Cave. It produced a stark contrast to darker pigments such as charcoal and red ochre, and was often used to highlight figures or to create a base layer. It was also used for body decoration. *Colour:* White.

Black Charcoal: Produced by burning wood, charcoal was readily available to prehistoric peoples and frequently used in cave art for outlining figures and for shading; examples seen at Chauvet and Lascaux caves. *Colour:* Rich black hue.

Manganese Dioxide: Collected from manganese-rich areas, this naturally occurring mineral could be ground into powder and mixed with water to create paint. It has been found in Palaeolithic cave paintings at Pech Merle and Lascaux. *Colour:* Dense, deep, black colour, slightly different in texture and sheen from charcoal.

Green Earth: Green earth pigments (celadonite and glauconite) were less common, but could be found included in certain mineral deposits across Europe. There is less evidence of green pigments in Palaeolithic cave art, but they were used in some instances, possibly for body adornment or in specific ritualistic contexts. *Colour:* soft, earthy green tones.

Bone Black: Produced by burning animal bones in a low-oxygen environment, resulting in a rich black pigment. Though less common than charcoal, bone black was occasionally used in prehistoric art for its finer texture and intense blackness, often for detailed work. *Colour:* rich, velvety black.

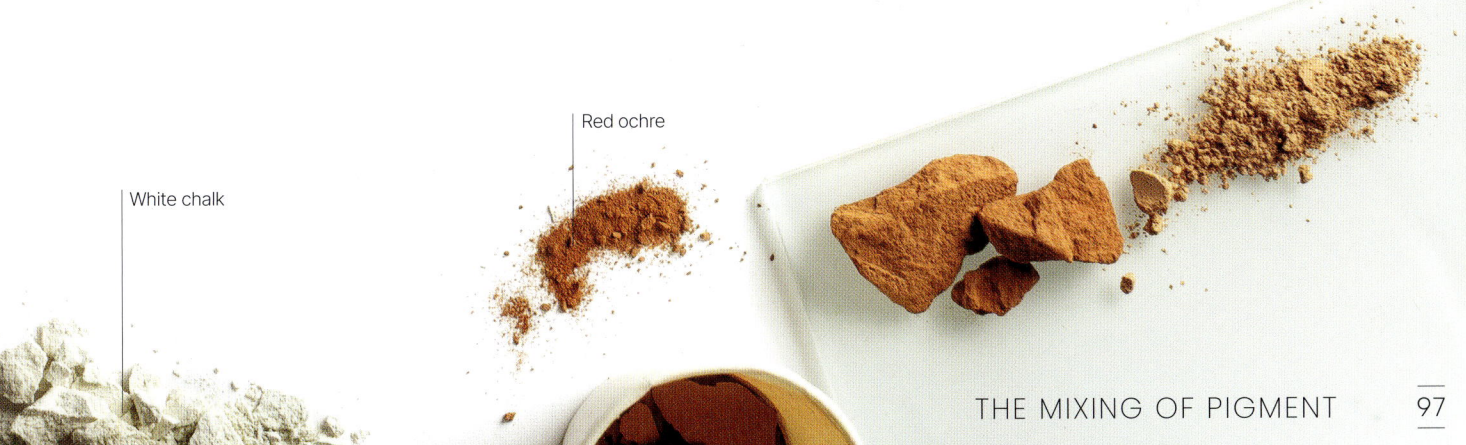

White chalk

Red ochre

THE MIXING OF PIGMENT

PIGMENT USE IN ROMAN BRITAIN

During the Roman era in Britain (43–410 CE), a variety of pigments were used in wall paintings, mosaics, and decorative arts, both locally sourced and imported from other parts of the empire. Thanks to their extensive trade networks the Romans had access to a wide range of pigments, and employed both mineral and organic pigments in their art.

The remains of Roman villas and of public baths in the UK often contain traces of these pigments, used to depict scenes of mythology, nature, and everyday life, such as those seen at Fishbourne Roman Palace (West Sussex). Roman mosaics in Britain, such as those found in Bignor Roman Villa (West Sussex) and Chedworth Roman Villa (Gloucestershire), used pigments and coloured stones to create detailed and vibrant floor decorations. Statues and architectural details were often painted using these pigments to enhance their appearance.

COLOURS IN ROMAN BRITAIN

Red Ochre: This iron oxide pigment (hematite) could be sourced locally in Britain. It was one of the most versatile pigments, commonly used in wall paintings and decorative art, for backgrounds, detailing, and shading. *Colour:* Shades of red, from deep reddish-brown to bright red.

Yellow Ochre: Another iron oxide pigment (goethite) found locally in Britain, similarly used to red ochre in wall paintings and frescoes, and it was often combined with other pigments to create different shades. *Colour:* Ranges from soft yellow to bright yellow to yellow-brown.

White Chalk: White pigments were often made from calcium carbonate, readily available in Britain, particularly in areas with chalk deposits such as the South Downs. White chalk was used in frescoes and as a base for mixing with other pigments to create lighter tones. *Colour:* Bright white with a yellow leaning.

Red Lead: A synthetic pigment made by heating lead. It was produced throughout the Roman Empire and used extensively in Britain. Red lead (minium) was often used in decorative wall paintings, particularly in higher-status buildings and public spaces where its bright colour was desired. *Colour:* Bright red to orange-red.

Vermilion: Brilliant red pigment made from cinnabar, a naturally occurring mineral (mercury sulfide). As cinnabar deposits were limited in Britain, the pigment was imported from other regions including Spain. Used in prestigious works of art, frescoes, and mosaics, typically in elite or religious buildings. *Colour:* Bright, vivid red.

Carbon Black: Produced by burning organic material, such as wood or bone, to create a deep black pigment; widely used for outlines, shading, and detailing in wall paintings and frescoes. *Colour:* Deep black.

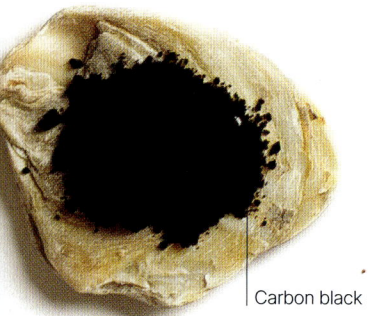

Carbon black

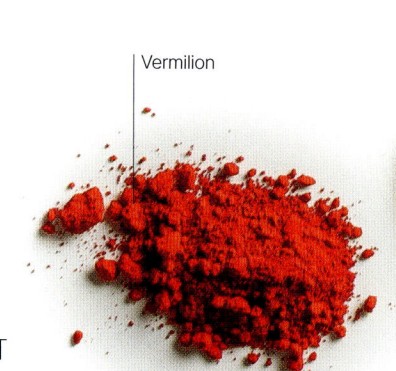

Vermilion

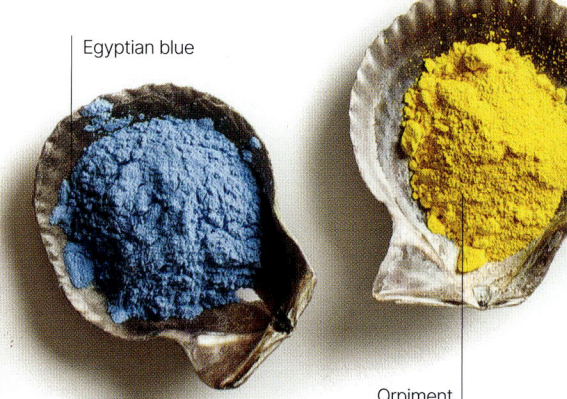

Egyptian blue

Orpiment

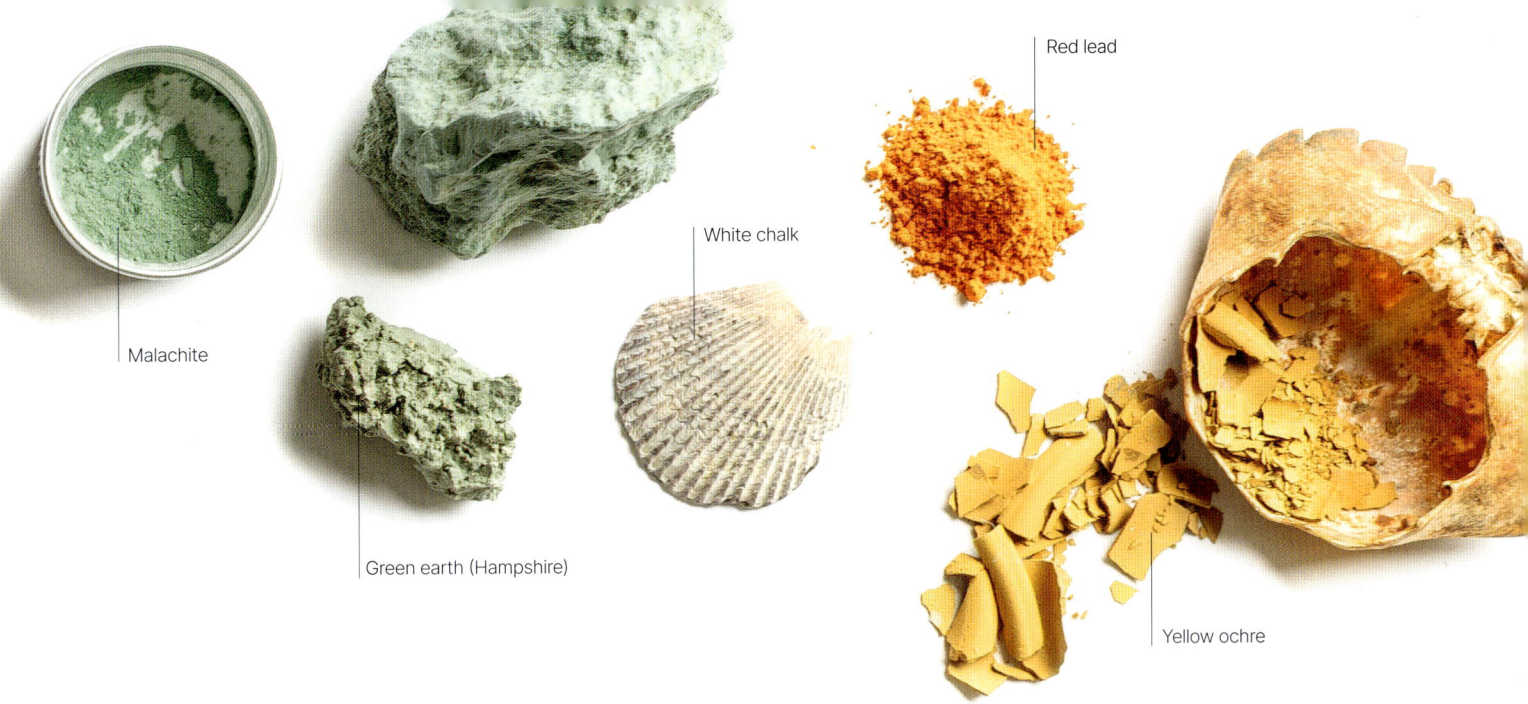

Malachite

Green earth (Hampshire)

White chalk

Red lead

Yellow ochre

Green Earth: Derived from minerals such as glauconite or celadonite, this was found in various parts of the Roman Empire, including Britain. Used in frescoes and mosaics, often to depict natural scenes or foliage. *Colour:* Soft, earthy green.

Egyptian Blue: One of the first synthetic pigments, made by heating a mixture of silica, lime, copper, and natron (a type of soda), resulting in calcium copper silicate. It was imported to Britain from other parts of the Roman Empire, e.g., Egypt. Used in high-status wall paintings and mosaics to depict clothing, skies, and decorative motifs. *Colour:* Bright blue with a slightly greenish tint.

Azurite: This copper carbonate mineral pigment was imported from copper-rich regions of the Roman Empire, such as the eastern Mediterranean. Used in frescoes and mosaics, especially for religious or high-status art. *Colour:* Deep blue.

Malachite: Another copper carbonate mineral pigment, also imported from copper-rich regions of the Roman Empire. Used in Roman wall paintings and mosaics to depict nature scenes, foliage, and sometimes decorative elements. *Colour:* Bright green.

Lead White: Produced by corroding lead metal in a vinegar environment, which created a bright white pigment (basic lead carbonate). Widely used across the Roman Empire, including in Britain. Used in frescoes and paintings, and as a base for mixing with other pigments to create lighter tones. *Colour:* Opaque, bright white.

Orpiment: As arsenic-rich minerals were less common in Britain, orpiment (arsenic sulfide) was imported from other regions of the Roman Empire. It was used for its bright, striking yellow hue, particularly in frescoes and mosaics. *Colour:* Bright yellow.

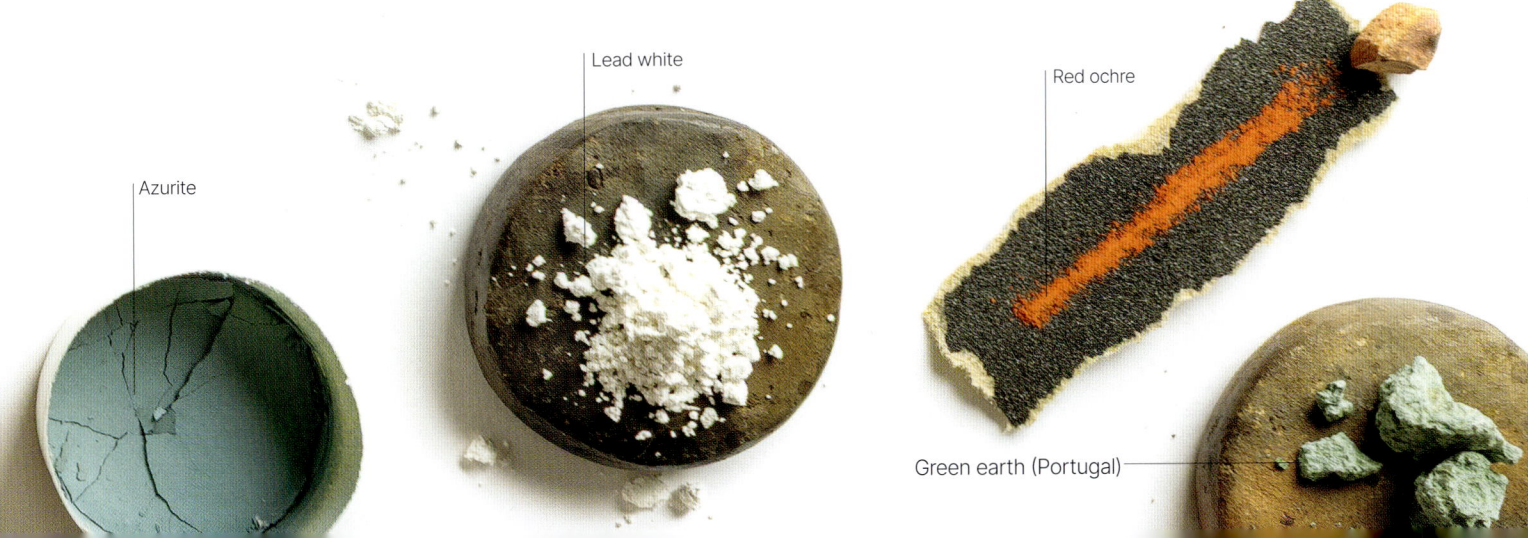

Azurite

Lead white

Red ochre

Green earth (Portugal)

THE ANGLO-SAXON (EARLY MEDIEVAL) COLOUR PALETTE

The Anglo-Saxons used a range of primarily natural pigments, either sourced locally from minerals, plants, and animal products, or imported from Europe. Some churches and domestic buildings were decorated with pigments applied to the plaster. Dyed fabrics and personal items were also common, especially among the nobility. Among the most famous examples of pigment use in this period, however, are illuminated manuscripts such as the Lindisfarne Gospels and the Book of Kells, showcasing intricate designs and vibrant colours.

Medieval illuminators preferred to use pure pigments rather than mixing them; they knew that some pigments were chemically incompatible with one another, causing discolouration, fading or even deterioration over time. Medieval art placed a high value on bright, vivid colours, and illuminators were often tasked with creating works that would catch the viewer's eye with intense luminosity. Instead of mixing pigments to create new colours, they often layered thin glazes of one colour over another to achieve depth and variation, allowing light to pass through the layers, creating a sense of richness and texture. This technique ensured that the properties of each pigment were preserved.

Pure pigments typically offered the richest and most saturated hues. The transparency and jewel-like appearance of azurite or malachite, for example, was best showcased alone. The aesthetic of medieval manuscripts was characterized by sharp contrasts and defined areas of colour, rather than the blending or gradation of tones that became more common in later artistic periods. The purity of pigments helped maintain these bold, separated colours, as well as their symbolic meanings.

High-quality pigments were expensive and difficult to obtain, especially for prized colours such as ultramarine (made from lapis lazuli), imported from Afghanistan. Illuminators were more likely to use these sparingly and in their purest form, to preserve this investment.

RECIPE: EGG TEMPERA

The most common medium for medieval manuscript illumination was egg tempera, where pigments were mixed with egg yolk as a binder.

INGREDIENTS

- One fresh egg
- A small amount of finely ground pigment powder
- Distilled water
- White vinegar or clove oil (optional)

METHOD

1. Carefully crack the egg and separate the yolk from the white. Roll the yolk in your palm to dry the membrane, then delicately pierce the membrane and allow the pure yolk to flow into a small container.

2. Place a small amount of pigment powder on a glass or ceramic palette. Add a few drops of distilled water to the pigment and mix with a palette knife to form a smooth paste.

3. Mix egg yolk and pigment paste in a 1:1 ratio, adjusting as needed to achieve the desired consistency. Thin the paint with a small amount of water to suit your painting style.

4. Egg tempera is best used fresh, but a drop of white vinegar or clove oil will extend its shelf life. Keep the paint in an airtight container in the fridge for up to two days, or until it starts to smell bad.

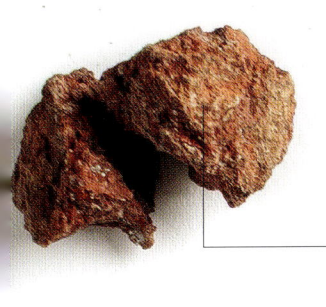

COLOURS IN MEDIEVAL BRITAIN

Red Ochre: This iron oxide pigment (hematite) was widely used in Anglo-Saxon manuscripts and murals, and for body decoration. *Colour:* A deep red to reddish-brown, depending on the concentration and purity of the iron oxide.

Yellow Ochre: This hydrated iron oxide (goethite) was commonly used in manuscripts and painting. *Colour:* Ranges from pale yellow to a more earthy brownish-yellow.

Green Earth: These pigments (made from minerals such as glauconite or celadonite), while not as common as ochre, were used in Anglo-Saxon art for their muted green hues. *Colour:* Soft, earthy green tones.

Red Lead: A synthetic pigment (minium) made by heating lead to create a vibrant colour, commonly used in manuscript illumination. *Colour:* Bright red to orange-red.

Woad: The leaves of this plant – native to Europe, including the UK – were processed to extract the blue dye, which was the primary source of this colour. Used in textiles and occasionally in painting. *Colour:* Medium blue, with earthy hue.

Verdigris: This is produced by exposing copper to acetic acid (vinegar), which creates a green patina. Commonly used for its brilliant green hue in manuscripts. *Colour:* Bright green, but highly reactive and prone to degradation over time.

Carbon Black: This pigment (charcoal), made from burned organic materials (e.g., wood or bone), was widely available in this period, used for drawing, outlining, and shading. *Colour:* Deep black.

Lead White: Made by corroding lead in the presence of vinegar and carbon dioxide, to create a white pigment often seen in Anglo-Saxon art. *Colour:* Bright, opaque white.

Ultramarine: Derived from lapis lazuli, this rare and expensive blue pigment was not common in Anglo-Saxon art, due to its cost, but was used for high-status manuscripts and religious works. *Colour:* Brilliant blue.

Organic Dyes: Organic dyes such as cochineal (red, from insects) and madder (red/orange, from plant roots) were also used, primarily for textile dyeing but occasionally for manuscript decoration. *Colour:* Vivid reds and oranges.

Chalk (Calcium Carbonate): This was another white pigment used in Anglo-Saxon art, particularly as a base layer, or for mixing with other pigments to lighten them. *Colour:* Soft white.

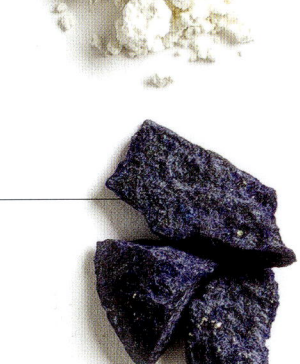

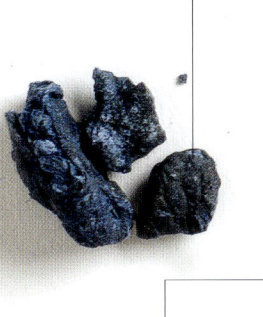

TUDOR COLOURS

The materials available for painting during the Tudor period in Britain (1485–1603) were in part similar to those used over the previous two centuries and across Europe. The primary palette featured a foundation of inorganic pigments derived from natural sources. Rich in these earth pigments, Britain exported red ochre at this time. Mineral-derived blues such as expensive lapis lazuli and azurite, alongside malachite, were also employed in easel painting. A rarer addition to the oil painting palette was the purple mineral fluorite, a calcium fluoride mineral sourced from Derbyshire. The rest of the Tudor palette was achieved through the more affordable processing of iron-rich clays to include yellow, orange, red, and black, with the addition of white chalk.

Red and yellow lakes, derived from dyes precipitated onto an alum and chalk substrate, were closely tied to the dyeing trade, using imported dyestuffs like kermes, madder, brazilwood, and weld. Woad, a major source of blue for dyeing, was also imported, and from it came the pigment indigo (or 'florey' as it is also known, from the Norman word for 'flower'), although superior indigo sources were also used. Lead-tin yellow (Type I) was linked to ceramic glaze production, and smalt, a cobalt-based blue, was associated with the glass industry – both were probably imported. The blue pigment verditer, which contained copper, could potentially be made locally due to the use of its components in other industries. Other imports included verdigris from Montpellier and high-quality vermilion from Antwerp, the primary trade hub for northwestern Europe in the 16th century.

English ports, especially London, regularly traded with ports in France, Germany, and the Netherlands. While London saw the largest influx of goods, other coastal and river ports were vital for distributing materials inland. Merchants across various London guilds sold painting materials, while grocers also stocked a broad but basic range. It was the apothecaries – who specialized more in pharmaceuticals – who were noted for selling pigments, possibly including higher quality colours such as ultramarine and crimson lakes.

A notable discovery was the use of colourless powdered glass in paint, frequently as a drying agent. The presence of powdered glass, particularly mixed with red lake pigment, in the paintings of Italian High Renaissance artist Raphael, was the first confirmation of this practice. The glass contains manganese, a known siccative (a substance added to a liquid to promote drying) for oils, making it ideal for use with transparent red glazes. Its function as a dryer is further evidenced by its hidden use in the mordant gilding of Raphael's *Ansidei Madonna*, where it would serve no visual purpose.

Archaeological studies reveal regional differences in glass composition: Italian paintings contain soda-lime glass, while northern European works show high-lime or mixed-alkali glass, reflecting local materials. Given its prevalence across Europe, it is probable that English artists also adopted powdered glass in 16th-century paintings. We know that 17th-century English sources recommend adding glass for this purpose.

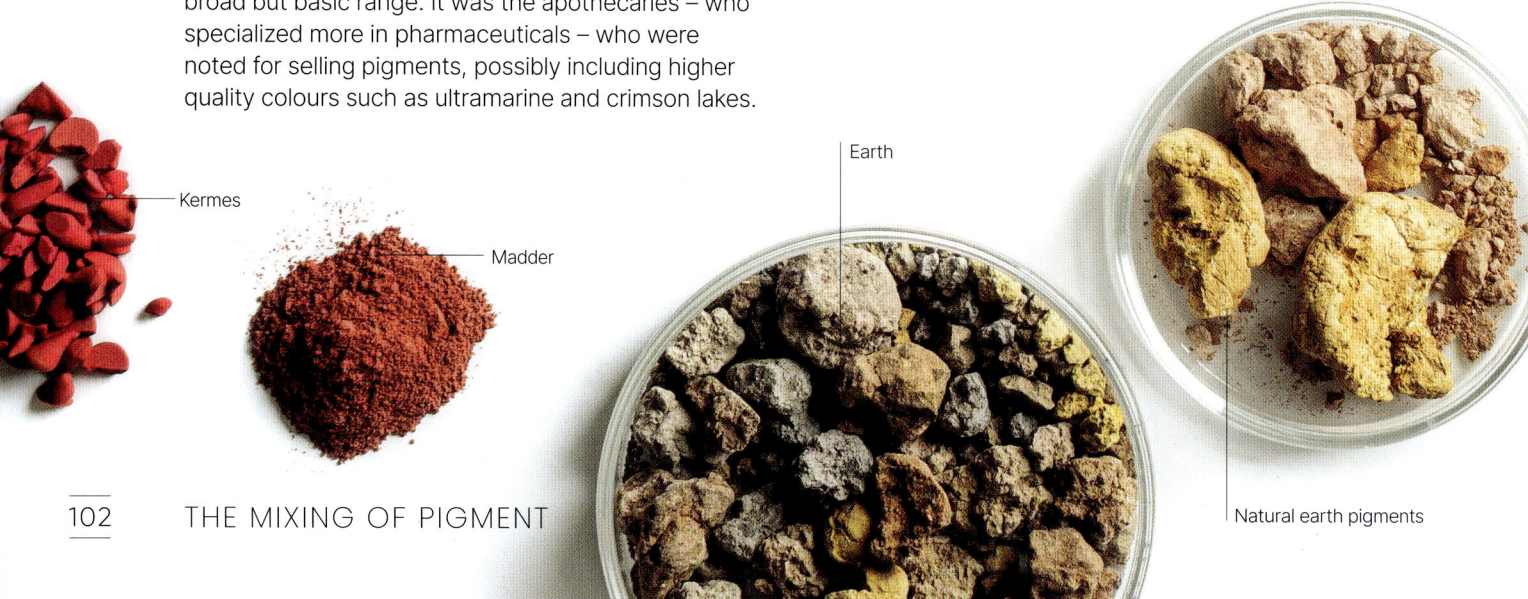

Kermes

Madder

Earth

Natural earth pigments

COLOURS IN TUDOR BRITAIN

Natural Earth Pigments: Used since prehistory, these are derived from naturally occurring mineral sources that contain iron and clay minerals called 'clay bodies' – specific masses or deposits of clay that have accumulated in localized areas. Bright yellow ochres from Shotover in Oxfordshire were used in local wall paints and for fine art applications. *Colour:* A full spectrum of colours from yellow, orange, red, to blue and purple, to black and white.

Blue Verditer: A bright blue pigment traditionally made by reacting copper salts (typically copper carbonate) in the presence of an alkali and water/air. It is a basic copper carbonate pigment, created through a chemical reaction that forms copper carbonate hydroxide. Tudor wall paintings from the late 1500s, now known as 'The Ledbury Painted Room', were discovered in 1988 in Ledbury, Herefordshire. This series of distemper paintings on lime features this synthetic blue pigment in its elaborate Elizabethan floral motifs. *Colour:* Blue and green.

Purple Fluorite Pigments: Derived from the mineral fluorite (calcium fluoride), which can exhibit a range of colours, including purple. To make the pigment, natural fluorite crystals are ground into a fine powder and treated to enhance their colour. The purple hues result from the presence of trace elements, such as yttrium or cerium, within the fluorite. Mined in Castleton in Derbyshire amongst other locations in northern Europe, it was used primarily by Flemish artists. *Colour:* Deep to light blue, leaning to purple.

Powdered Glass: When finely ground and mixed into oil paint, powdered glass could enhance the paint's texture, sheen, and luminosity. It helped to increase the viscosity of the paint, making colours go further. Additionally, powdered glass could create a smooth, glossy finish once dried, giving the painted surface a reflective quality. *Colour:* Colourless to white depending on particle size.

Also continuing in popularity in the Tudor palette were **lapis lazuli**, **vermilion/cinnabar**, **azurite**, **malachite**, **red dyes/lakes**, and lead colours (**tin/orpiment**).

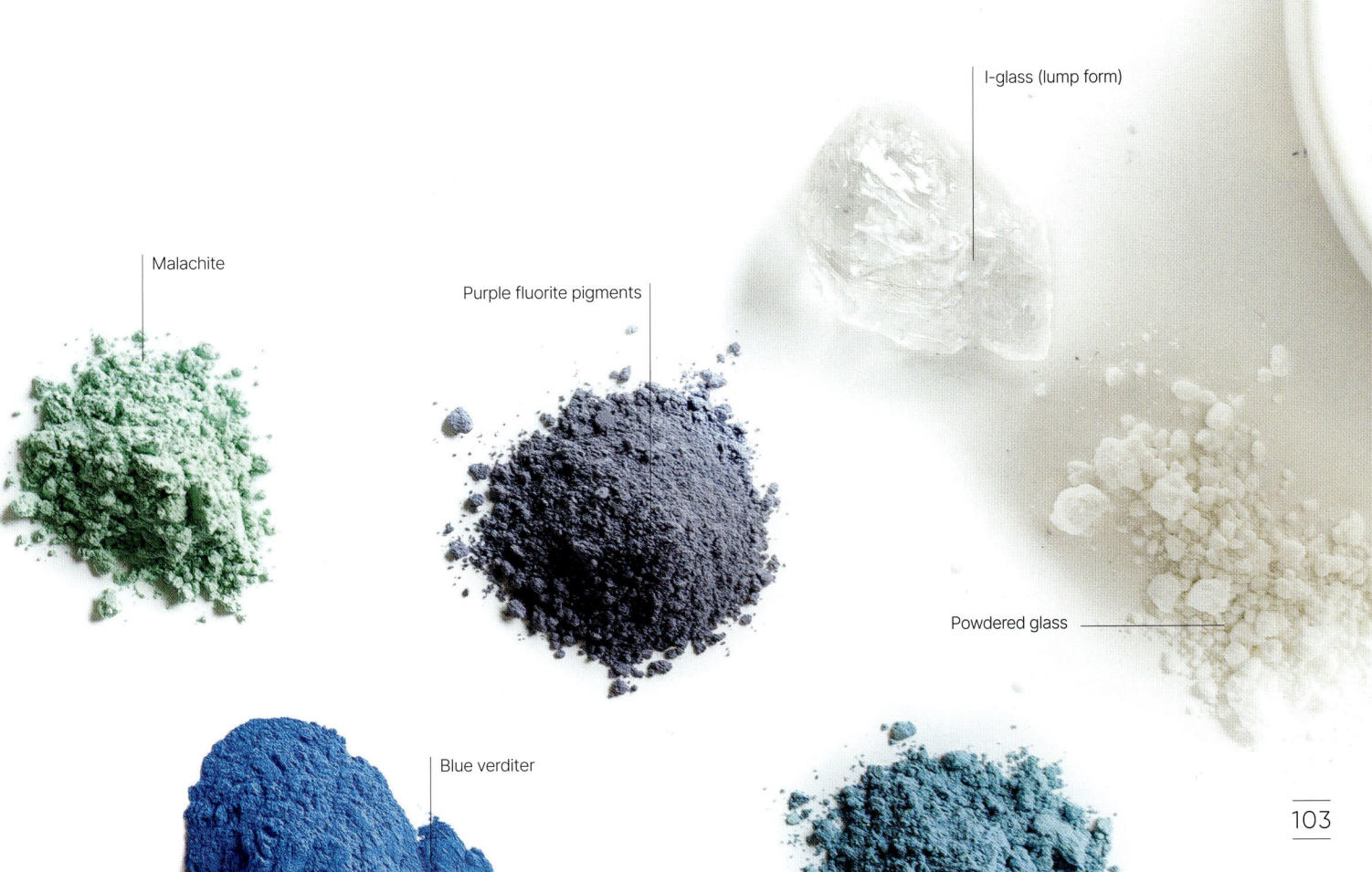

Malachite

Purple fluorite pigments

I-glass (lump form)

Powdered glass

Blue verditer

APPROPRIATED COLOURS

In Victorian Britain (1837–1901), the expansion of the British Empire unlocked questionable access to exotic dyes, pigments, and raw materials that transformed the British visual landscape. These vibrant colours, sourced directly from colonies or along British-controlled trade routes, became emblems of wealth, sophistication, and status, reflecting the vast economic and political reach of the empire.

The rich hues filling the textiles and artworks of this period underscored the nation's dependence on natural resources drawn from its colonies, driving not only industrial growth but also a flourishing demand for new alluring colours. Each shade embodied Britain's pursuit of progress, innovation, and luxury – yet also reveals the environmental strains and ethical costs bound to colonial practices. Thus, these colours became potent symbols of Britain's prosperity and industrial advances, and of the complex and troubling legacies of empire.

Our use of these colours today, with their chemical compounds often hidden on labels, conceals histories intricately linked to the sourcing and production of the pigments, which are frequently disconnected from their original origins. Untangling these intertwined narratives allows us to restore agency to the traditional knowledge holders and the organisms that played a role in their creation.

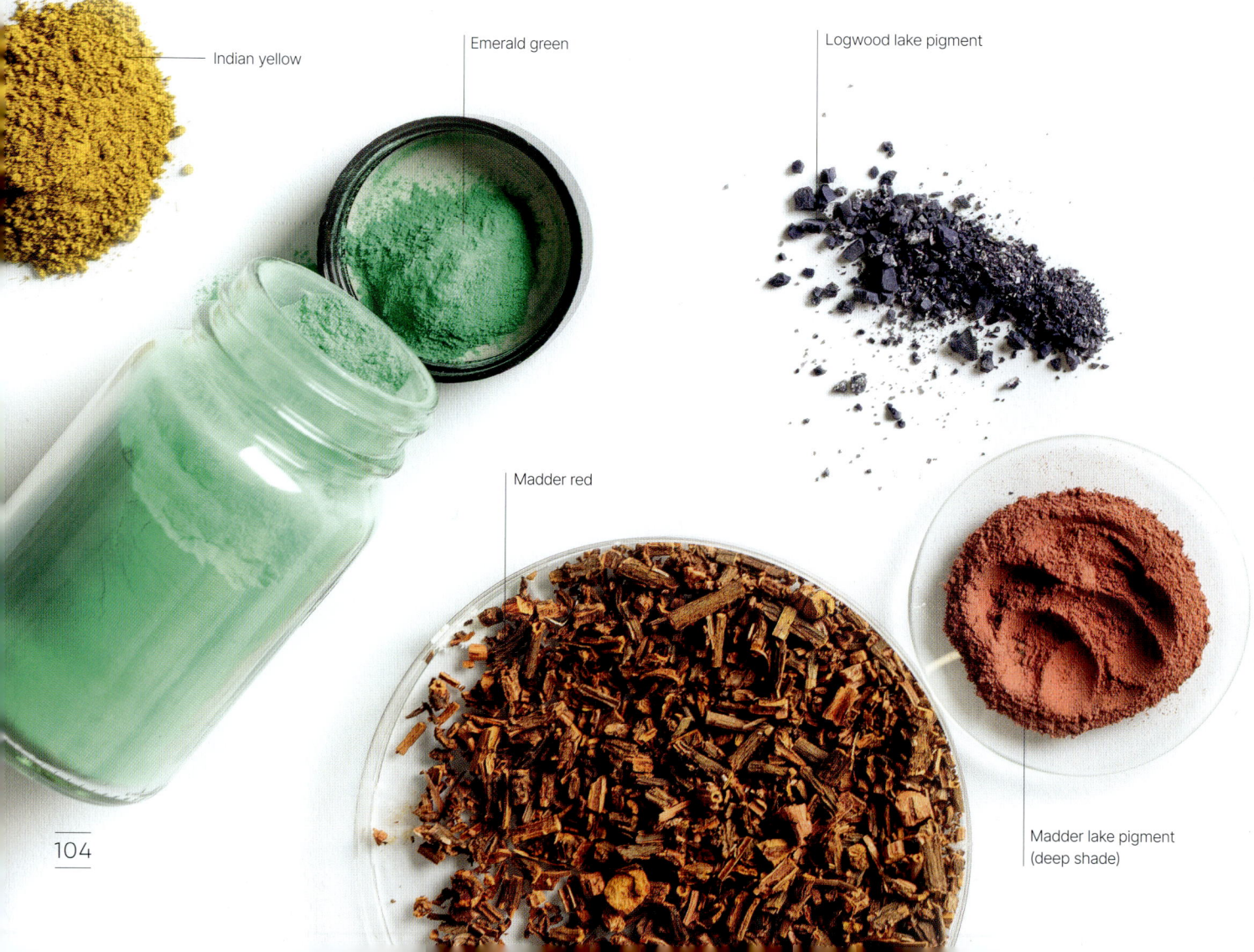

Indian yellow

Emerald green

Logwood lake pigment

Madder red

Madder lake pigment (deep shade)

PROBLEMATIC PALETTES OF 'EMPIRE'

Indian Yellow: A vibrant yellow pigment originally made from the urine of cows fed on a diet of mango leaves. The urine was evaporated to produce a concentrated, golden-coloured substance that was dried and ground into a powder. Used since the 15th century, it gained prominence in Europe, especially after the British occupation of India. English Romantic painter J.M.W. Turner used it to great effect in rendering light. Ethical concerns led to its decline, and synthetic alternatives (tartrazine) now replicate its bright, transparent yellow hue. *Colours:* Warm yellow, diluted by a medium; brown mustard, undiluted.

Emerald Green: A synthetic pigment made from copper arsenite. Affordable copper was sourced through colonial trade routes via South Africa, Chile, and Australia. It is created by reacting arsenic trioxide with copper salts, typically copper acetate, in the presence of an alkaline substance. The resulting colour was widely used in art and decor (wallpapers, paints, and textiles) during the 19th century. Highly toxic due to its arsenic content, this led to its decline in use. *Colour:* Bright green.

Logwood, Brazilwood, and Sappanwood: These natural dye sources were derived from the heartwoods of trees native to tropical regions, with each producing vibrant hues that were highly valued during the colonial period. Logwood, sourced from *Haematoxylum campechianum* in Central America, yields deep purple to black dyes creating rich, dark colours in textiles and inks. Brazilwood, from *Caesalpinia echinata* in Brazil, produces a red dye. Prized for its brilliant colour, it was used widely in textiles and art. Sappanwood, from the *Caesalpinia sappan* tree in South-east Asia, yields a red to

orange dye. While all three woods were significant in the colonial dye trade, they were ultimately replaced by cheaper, synthetic alternatives as the industrial era progressed. *Colours:* Purple, black, red, orange-red.

Indigo: *Indigofera tinctoria* is native to India, as well as other regions in South-east Asia and Africa. It has been cultivated in India for centuries, if not millennia, and played a significant role in the country's historical textile industry and global trade. The Indian subcontinent's climate, particularly in tropical and subtropical regions, is well-suited to its growth, making it one of the primary sources of natural indigo dye. Its historical importance in India is evident in its integration into traditional practices, the export of indigo dye, and its influence on colonial economies. Indian-grown indigo was utilized widely by the Pre-Raphaelite brotherhood, alongside William Morris, who also sourced it's sister plant, woad, closer to home. *Colour:* Blue.

Madder Red: The madder plant (*Rubia tinctorum*) was cultivated in the British colonies in India, and leveraged by the colonialists to exploit and dominate the textile dyeing and trade industries. Madder dyes were crucial in producing Indian chintz and calico textiles, renowned for their intricate red patterns, and highly sought-after in European markets. While Indian farmers and artisans received minimal profits, British industries benefited immensely, disrupting traditional dyeing practices and further entrenching India's economic dependency on colonial policies. *Colour:* Red.

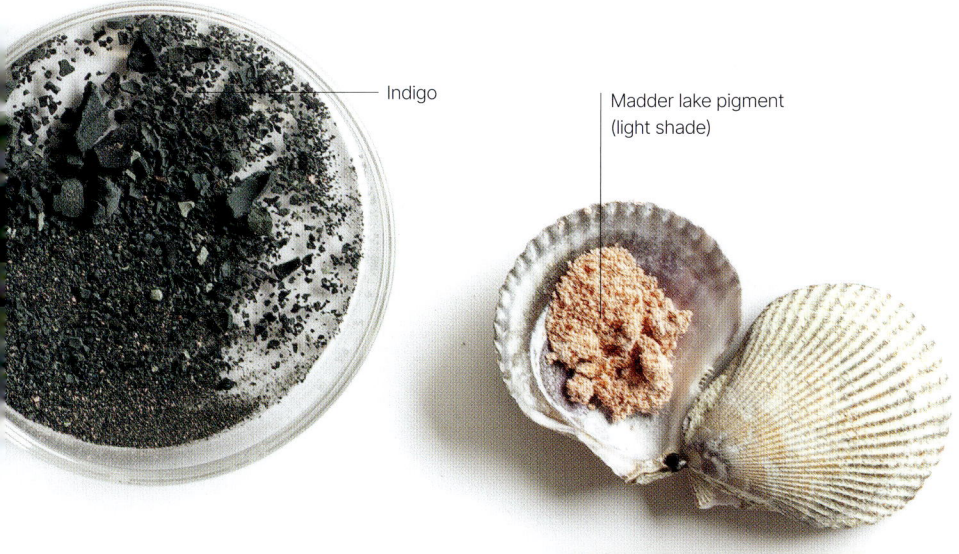

Indigo

Madder lake pigment (light shade)

COCHINEAL AND SPANISH COLONIALISM

Colonial trade was not, of course, solely a British occupation. Following the Spanish expedition to Mexico in 1518, commanded by Hernando Cortés, cochineal dye was exported to Europe. It had been used by the Aztec and Maya peoples of North and Central America for many centuries.

Highly valued for its deep and long-lasting red colour, it was considered a superior dye to madder. The Spanish kept the source of the cochineal secret and for nearly 200 years it was thought be a plant seed. Cochineal is, in fact, made from various species of scale insects (*Dactylopius*), small bugs fairly closely related to aphids or cicadas. Contrary to popular belief, they are not beetles. They feed on the *Opuntia* cactus.

Spain's powerful monopoly over the commodity included prohibiting production of cochineal outside of the state of Oaxaca, to which it is native. After Mexican independence in 1821, however, Spain urgently needed to find a new location for their cash crop and began to cultivate cochineal in the Canaries.

BASIC METHOD FOR PRODUCING COCHINEAL PIGMENTS

The red dye is extracted from the insects using water and heat, and then filtered. A substrate, alum in the case of the recipe provided (see Recipe: Contemporary Cochineal), is added to the decoction as the material for the dye to attach to. Potassium carbonate is then added to 'fix' the dye to the alum. After this chemical reaction has taken place, the particles precipitate out of the solution and the liquid is filtered off. The remaining pigment is washed several times so that any remaining unreacted ingredients are disposed of. The pigment is left to dry, and then it is ground using a pestle and mortar.

Three key cochineal pigments are:

Cochineal Crimson: A deep pink-red pigment made from tin chloride, alum, and potassium carbonate).
Cochineal Pink: A soft candy-coloured pigment made from alum, potassium carbonate, and chalk.
Cochineal Purple-Red: A mid-tone purple made from alum and potassium carbonate with alum added second.

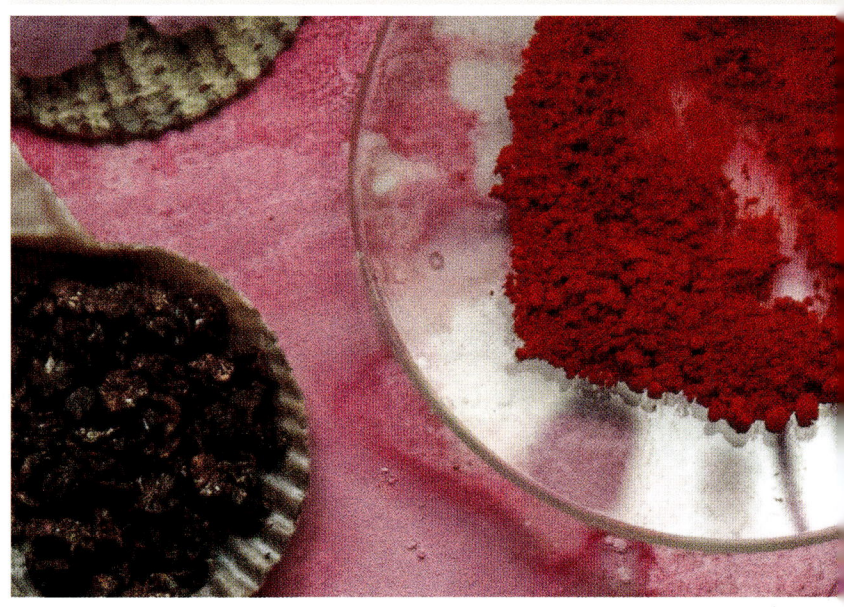

RECIPE: CONTEMPORARY COCHINEAL

Cochineal and carmine are general colour names for a range of lake pigments, encompassing shades of red, pink, and purple, derived from over 10 individual species of scale insects. Their potentially unsettling origins are somewhat alleviated by the fact that the insects were often collected and killed as agricultural pests in South America and Spain. However, dedicated commercial production also exists for these vibrant and historically significant colours. Although technically a non-toxic pigment, do not use on the body or consume. Not suitable for vegans and vegetarians.

INGREDIENTS

- Alum (potassium aluminium sulfate), 30g
- Potassium carbonate, 10g
- Cochineal insects, 5g
- Distilled water

EQUIPMENT

- PPE – respirator (P3 rating)
- Heat-resistant glass beakers, two 300ml and one 750ml
- Kitchen pan
- Thermometer
- Stirrer
- Coffee filter papers
- Funnel
- Pestle and mortar
- Sieve (40-mesh)
- Muslin cloth

METHOD

PREPARING THE SOLUTIONS

1. Dissolve the alum in 200ml of boiling distilled water.

2. In a separate container, dissolve the potassium carbonate in 100ml of boiling distilled water.

3. Place the cochineal in a piece of muslin, gather the corners, and tie securely with twine. Simmer the pouch in distilled water for 10 minutes.

4. Using a thermometer check the solutions are approx. 50°C (122°F), adding a little cold distilled water if necessary.

COMBINING THE SOLUTIONS

1. Add the alum solution to the cochineal solution in the 750ml beaker and mix well.

2. Slowly add potassium carbonate solution to the combined alum-cochineal solution and mix well, until foaming no longer occurs.

3. Wait for the pigment to settle out of the solution, then pour off the water, leaving pigment slurry in the bottom of the beaker.

CLEANING THE PIGMENT

1. Wait for pigment to settle and pour away excess water. Pour the remaining pigment slurry into a coffee filter paper placed in a funnel.

2. Add more water to the filter paper once the original water has drawn through to wash the pigment. Gently mix and repeat as necessary.

DRYING THE PIGMENT

1. Once all remaining water has been filtered out, rip open the filter paper at the seams to place it flat on an even surface, and spread the wet pigment out onto it thinly.

2. Once thoroughly dry, grind the pigment using a pestle and mortar until you have reduced it to a 40-mesh powder, using a 40-mesh sieve to check.

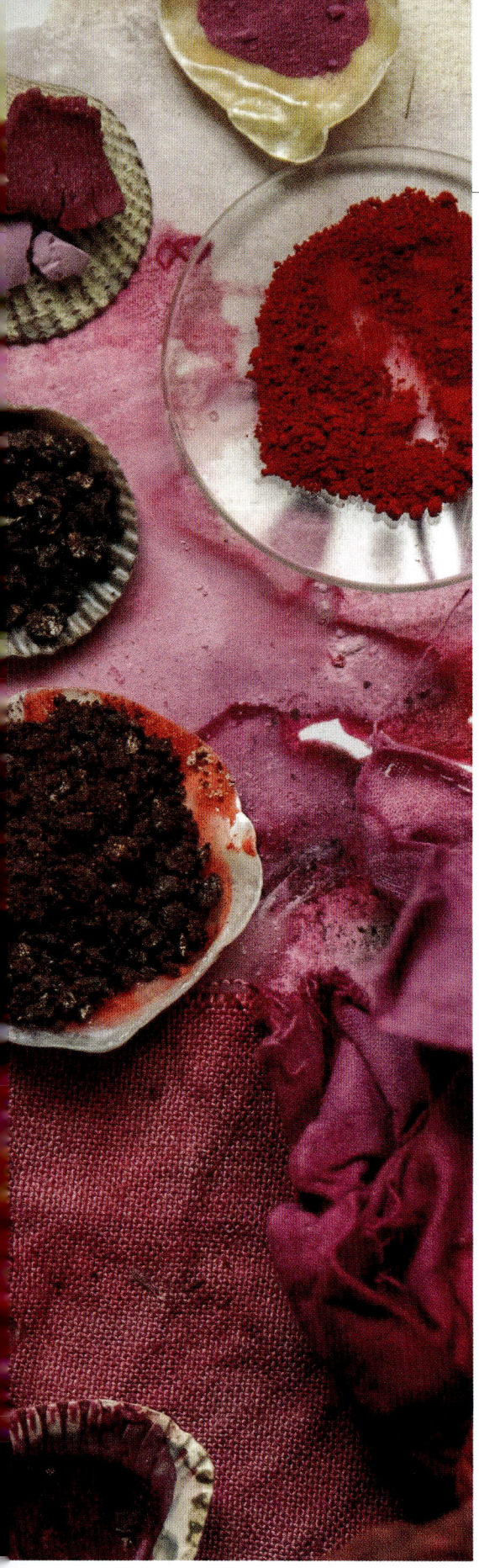

PIGMENT SUMMARY

Pigment Name: Cochineal (from most widely used *Dactylopius coccus* species)

Pigment Nomenclature: Carmine, crimson lake, Mexican red, Nocheztli, cochinilla, grana

Pigment Index Number: NR 4

Colours: Vibrant red through crimson and purple

Colourant & Chemical Formula: Carminic acid (main dye colourant used) $C_{22}H_{20}O_{13}$

Classification: Natural organic (from animal origin)

Pigment Description: Derived from the dried bodies of pregnant female cochineal insects, which are parasitic on specific cacti, primarily in Central and South America. It is processed by extracting carminic acid, which is then precipitated onto an alum base to create a lake pigment. Known for its rich, red hues, high tinting strength, and excellent transparency, although it is moderately fugitive under certain conditions

Conservation Issues: Sensitive to light and environmental pH. Prolonged light exposure can lead to fading, particularly in alkaline environments. May also react with other pigments or binders, causing discolouration

Particle Morphology: Irregular, soft particles, often amorphous or flake-like, depending on the preparation method

Dates in Use: Pre-Columbian (*c.* 2000 BCE in Mesoamerica); widely adopted in Europe from 16th century onwards

Density: Approx. 1.5g/cm³

Hardness: Variable; generally 2 on the Mohs scale

Oil Absorption Ratio: Moderate; 40–50g of oil per 100g of pigment

Refractive Index: Approx. 1.52

Health and Safety: Generally non-toxic and safe to handle. However, some individuals may experience allergic reactions or sensitivities, particularly when it is ingested or used in cosmetic formulations.

Environmental Impact: Cochineal cultivation, while resulting in the large-scale death of insects, also helps reduce waste, as some species are considered pests that feed on plants

INSECT DYES

Several species of adult female parasitic insects can be used to create brightly coloured organic dyes and lake pigments. Featured here are the most common. 'New World' refers to the European exploration of the Americas in the 15th century, while 'Old World' refers to historic availability in Europe, Africa and Europe before that time.

'NEW WORLD' COCHINEAL

COCHINEAL (*DACTYLOPIUS COCCUS*)

Colourant: Carminic acid (20% by weight)

Lake Colour: Red, pink, purple, red-purple

Origin: Mexico and Peruvian Andes before being brought to Canary Islands and Spain

Host Plant: Prickly pear or Barbary fig cactus (*Opuntia ficus-indica*) and usual host for farming methods

Date of Use: First arrived in Spain in 1523; used by Indigenous Peoples of Mexico and Peru for the last 1,500 years

Key Details: More efficient to use than 'Old World' species. A wild species, *Grana silvestre*, can be found on *Opuntia* cactus; however, as smaller, more would be needed for best dyeing results.

'OLD WORLD' COCHINEAL

KERMES (*KERMES VERMILIO*)

Colourant: Red anthraquinone kermesic acid + some yellow flavokermesic acid

Lake Colour: Scarlet red

Origin: Spain, southern France, North Africa

Host Plant: Kermes oak (*Quercus coccifera*)

Date of Use: Late 14th century to 16th century as a lake pigment in oil

Key Details: Important European dye before arrival of 'New World' cochineal in 16th century; insect is now very rare

LAC PRODUCED BY SEVERAL SPECIES: (*KERRIA LACCA; LACCIFER LACCA; TACHARDIA LACCA*)

Colourant: Anthraquinone, laccaic acid, erythrolaccin, flavokermesic acid (as found in kermes)

Origin: India, China, Sri Lanka, Burma, Pakistan

Host Plant: Bastard teak (*Butea monosperma*)

Date of Use: 13th century, maybe earlier; in European painting as lake pigments and for 17th century Coptic textiles

Key Details: Made from stick lac (sticky brown resinous secretions) from shellac. An important historical colourant, the resin is washed in dilute solution of sodium carbonate to extract dyestuff/

POLISH COCHINEAL (*PORPHYROPHORA POLONICA*)

Colourant: Carminic acid, 20–25% kermesic and flavokermesic acids

Lake Colour: Red

Origin: Central and Eastern Europe: Lithuania, Poland, Germany, the Czech Republic, Hungary, Ukraine, Russia, Mongolia

Host Plant: Perennial knawel (*Scleranthus perennis*)

Date of Use: Used as a red dye from 15th–19th centuries. Its use declined in the 1800s due to the importation of *Dactylopius coccus*, which provided a more efficient alternative

Key Details: This species is part of the Porphyrophora family of scale insects, which includes several species known for their dye-producing capabilities, including *Porphyrophora hamelii* (Armenian cochineal). Polish cochineal produced rich shades of red, crimson, and scarlet, with carminic acid as its main dye compound, similar to Mexican cochineal. Highly valued for royal and ecclesiastical garments, it played a significant role in the economy of the Polish-Lithuanian Commonwealth.

THE COLOUR PALETTE
OF THE BAUHAUS SCHOOL

Founded in 1919 by Walter Gropius in Weimar, Germany, the Bauhaus School was a revolutionary art and design school that emphasized the integration of art, craft, and technology. The Bauhaus movement was marked by a unique approach to colour and pigment use, combining traditional artistic techniques with new industrial materials.

Bauhaus artists often employed simplified colour palettes focused on primary colours (red, blue, yellow) and secondary colours (green, orange, violet) as part of their modernist design philosophy. These colours were used in their purest form to achieve bold, geometric designs and were influenced by colour theory. Bauhaus artists placed a strong emphasis on the psychological and emotional effects of colour. They often experimented with colour juxtaposition and transparency, and colour gradients, which influenced their choice of pigments. The school's approach to colour was both analytical and expressive, using pigments to explore contrasts, saturation, and balance.

Johannes Itten, one of the early instructors at the Bauhaus, developed a colour wheel based on his colour theory, and this became a central part of the Bauhaus curriculum. Paul Klee and Wassily Kandinsky also contributed to the understanding and use of colour as a tool for emotional expression and compositional harmony.

Artists at the Bauhaus embraced organic pigments, such as madder lake and indigo, as well as more modern synthetic pigments. Their use of industrial paints and enamels reflected the Bauhaus's commitment to blending fine art with functional design and mass production. By this time, the chemical industry had developed a wide array of synthetic pigments that were widely available. Cadmium red and cadmium yellow were bright and durable – firm favourites for their strong colour intensity and good lightfastness. Phthalo blue and phthalo green were created from phthalocyanine, and both were prized for their vivid colour and stability. Titanium white was a highly opaque and bright white pigment that became the dominant white pigment of the 20th century, replacing lead white due to its non-toxic nature.

At the Bauhaus, the combination of historic pigments (such as earth pigments) and newly industrially produced ones, paired with the use of trade-standard paint binders, highlights how the cultural and technological context of the time shaped material choices.

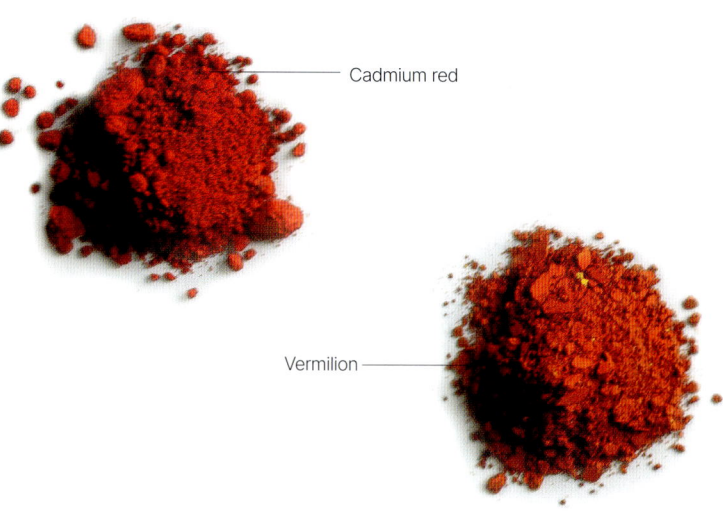

Cadmium red

Vermilion

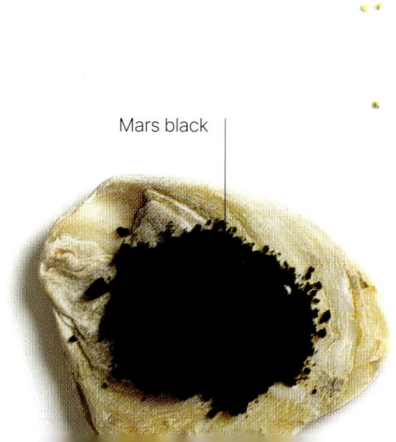

Mars black

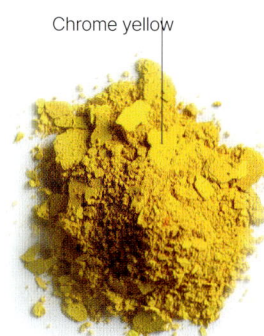

Chrome yellow

PRIMARY COLOURS AND PIGMENTS USED AT THE BAUHAUS

Cadmium Red: This synthetic pigment became widely available in the early 20th century. It was favoured by Bauhaus artists for its intensity and durability. It has strong covering power and excellent permanence, making it ideal for murals and paintings, and even industrial applications such as enamel. *Colour:* Bright, warm red with excellent opacity and lightfastness.

Vermilion: Although expensive and less commonly used by the time of the Bauhaus, vermilion was a traditional pigment made from mercury sulfide. Some artists may still have used it for specific applications, though synthetic reds were more common by this period. *Colour:* Brilliant, slightly cooler red hue, known for its intensity.

Red Lead: Another traditional red pigment that was still in use at the time, though it was more associated with industrial paints. Red lead (minium) has a long history of use in European art. *Colour:* Bright, orange-red colour.

Ultramarine Blue: The synthetic form of ultramarine blue, developed in the early 19th century; widely used by Bauhaus artists. It was much more affordable than the original natural ultramarine made from lapis lazuli, and it provided a deep, vibrant colour that was highly valued for its intensity and its excellent lightfastness and stability. *Colour:* Rich, transparent blue.

Phthalo Blue: This synthetic pigment (also called phthalocyanine blue) was developed in the early 20th century and quickly became a popular choice for artists, including those at the Bauhaus. It has excellent lightfastness and permanence. *Colour:* A powerful, highly chromatic blue that can range from cool to warm depending on how its mixed or used.

Cobalt Blue: This synthetic pigment, developed in the 19th century, was also used by Bauhaus artists. More expensive than ultramarine, it was valued for its subtlety and brightness. It has a strong tinting strength and is highly stable, making it suitable for both artistic and industrial applications. *Colour:* Bright, cool blue with good opacity and permanence.

Cadmium Yellow: This synthetic pigment became the dominant yellow pigment of the early 20th century. Bauhaus artists favoured it for its bright, rich hue and excellent opacity. It is highly durable, lightfast, and opaque, making it ideal for creating solid fields of colour. *Colour:* Ranging from light lemony shades to deep, warm yellows.

Chrome Yellow: Still in use at this time, although it was gradually replaced by cadmium yellow due to concerns over its toxicity. Chrome yellow (lead chromate) provided a brilliant colour but had lower lightfastness and could degrade in the presence of sulfur or pollution. *Colour:* Bright yellow.

Lead-Tin Yellow: This traditional pigment was still occasionally seen in the early 20th century, but due to its toxicity it was phased out in favour of cadmium yellow. *Colour:* Warm, opaque yellow.

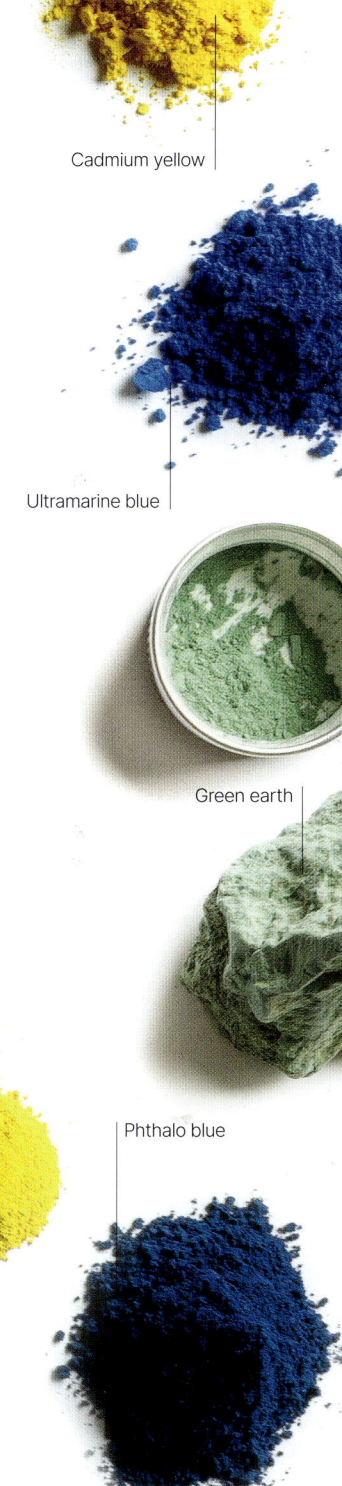

Red lead

Cadmium yellow

Ultramarine blue

Green earth

Lead-tin yellow

Phthalo blue

Cobalt blue

PIGMENT INCOMPATIBILITIES

Most modern synthetic pigments in commercial paint ranges offer artists exceptional stability and lightfastness. However, historically authentic, handmade pigments present a different story. Often crafted from raw materials with variable purity and complexity, sometimes with unpredictable compositions, these can produce unusual and sometimes unintended effects.

Throughout history, it became apparent that certain pigments reacted poorly when mixed, leading to discolouration or degradation in artworks. This issue is particularly notable in artworks created before the advent of modern synthetic pigments, as historical pigments were often composed of minerals or organic compounds that could interact in unstable ways. These incompatibilities underscore the challenges artists faced when selecting and mixing pigments. Without a deep understanding of chemistry, they had to rely on practical knowledge, observation, and caution to avoid unwanted reactions in their artwork. The legacy of these challenges is still visible today, as conservators work to stabilize and restore historical pieces impacted by these reactive combinations.

VERDIGRIS

A green pigment derived from copper that was popular in the Middle Ages. It is highly reactive and unstable when combined with sulfur-containing pigments, such as synthetic ultramarine, natural lapis lazuli pigments, or vermilion (mercury sulfide). When combined, these pigments form copper sulfides, leading to discolouration and often darkening the original colours. Copper compounds are also known to be reactive over time, causing these mixtures to lose vibrancy or change colour significantly. Due to this instability, artists became cautious in mixing verdigris with other pigments, particularly in oil paintings where chemical interactions are more intense. Verdigris compounds are acidic and therefore can have a detrimental affect on substrates if not isolated thoroughly.

AZURITE

Due to its tendency to oxidize on contact with air and water, this is an unstable pigment. It transforms into malachite, which turns blue areas green or darker over time. It is especially sensitive to humidity and pH changes, degrading in moist or acidic conditions. Also, azurite does not bind well in oil and may react with sulfur-based pigments, leading to further darkening.

LEAD-BASED PIGMENTS

Lead white and red lead (also known as minium) are highly reactive with certain other pigments, especially copper-based pigments such as malachite and azurite. When mixed, the lead reacts with copper compounds to form lead sulfates or lead chlorides, which can lead to blackening or browning of the paint. Lead white was a staple pigment in historical artworks, prized for its opacity and warmth; however, in combination with copper-based greens or blues, this pigment could compromise the durability and appearance of a painting over time. Conversely, it was added to blue verditer to help with its stability over time!

ORPIMENT

This vibrant yellow pigment, made from arsenic sulfide, is another problematic pigment due to its reactivity and toxicity. When mixed with lead-based pigments such as lead white or red lead, orpiment can form lead arsenates, which cause degradation in both pigments. Orpiment was also unstable in light and has a high tendency to darken, further limiting its use in mixtures and often contributing to issues in historical works.

Madder lake pigment (here as a watercolour) can lighten over time if left in direct sunlight

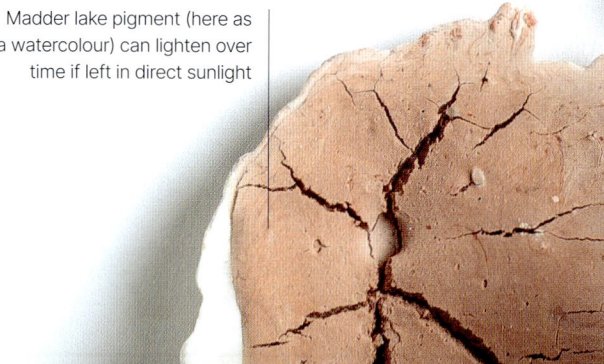

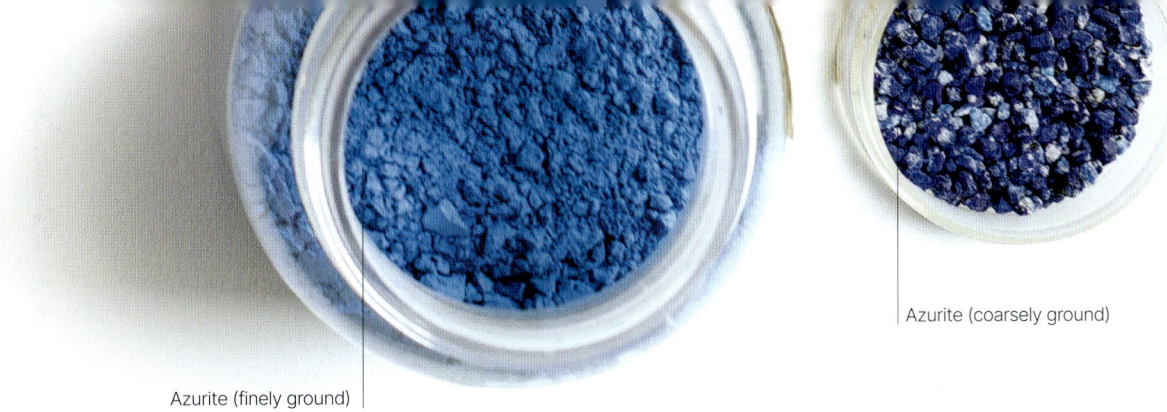

Azurite (finely ground)

Azurite (coarsely ground)

IRON OXIDE PIGMENTS

Red ochre and yellow ochre are susceptible to chemical changes when mixed with strongly alkaline pigments such as smalt or lime-based whites. The alkali can react with the iron oxides, potentially leading to colour changes (loss of vibrancy), even causing flaking (loss of pigment adhesion). This was particularly challenging in frescoes, where the lime binder could interact unpredictably with the iron oxides in natural earth colours.

PRUSSIAN BLUE

This early synthetic blue pigment, made of iron compounds, is known to be sensitive to alkaline conditions. When used in combination with lime white (as in fresco painting), Prussian blue can fade or shift to a greenish hue. This reactivity is due to the instability of the iron compounds in an alkaline environment, making it unsuitable for certain applications, despite its popularity as a deep, intense blue. (It can also turn black if kept out of sunlight!)

LAKE PIGMENTS

Made by precipitating dyes onto an inert substrate (often a metallic salt or alum), these are prone to instability when mixed with metallic pigments, such as lead-based or copper-based colours. The organic dyes in lake pigments can fade rapidly or change colour, especially in high-humidity conditions or under prolonged exposure to light. This sensitivity often made lake pigments less reliable for long-lasting artworks, although they were valued for their bright, intense colours. In many Renaissance paintings, generations later, flesh colours often appear as greenish tones. This is because the warm tones for human faces, etc., were created with red lake pigments, derived from organic dyes like madder or cochineal. Over time, these pigments faded, leaving behind the underlying greenish or bluish pigments such as copper-based greens (including verdigris or azurite) or green earths. If washed insufficiently during their manufacture, these pigments can fade quickly due to extremes in pH created from leftover reagents.

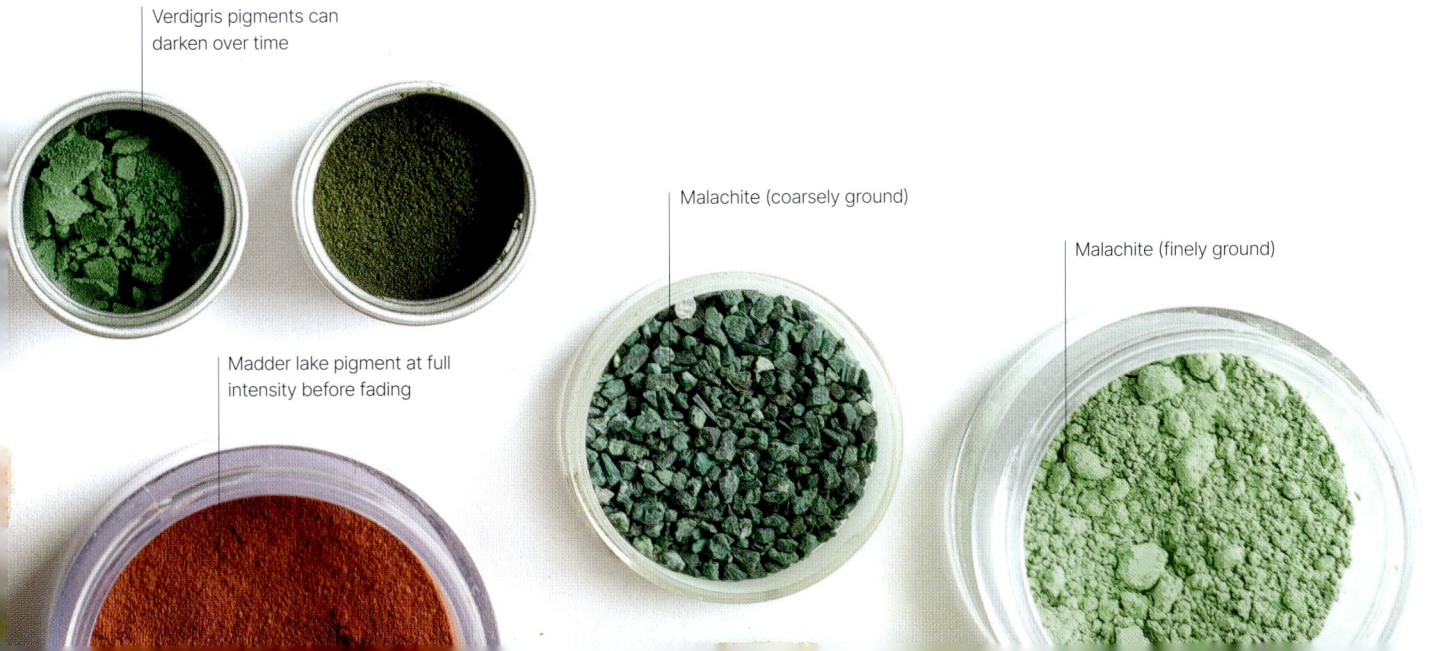

Verdigris pigments can darken over time

Madder lake pigment at full intensity before fading

Malachite (coarsely ground)

Malachite (finely ground)

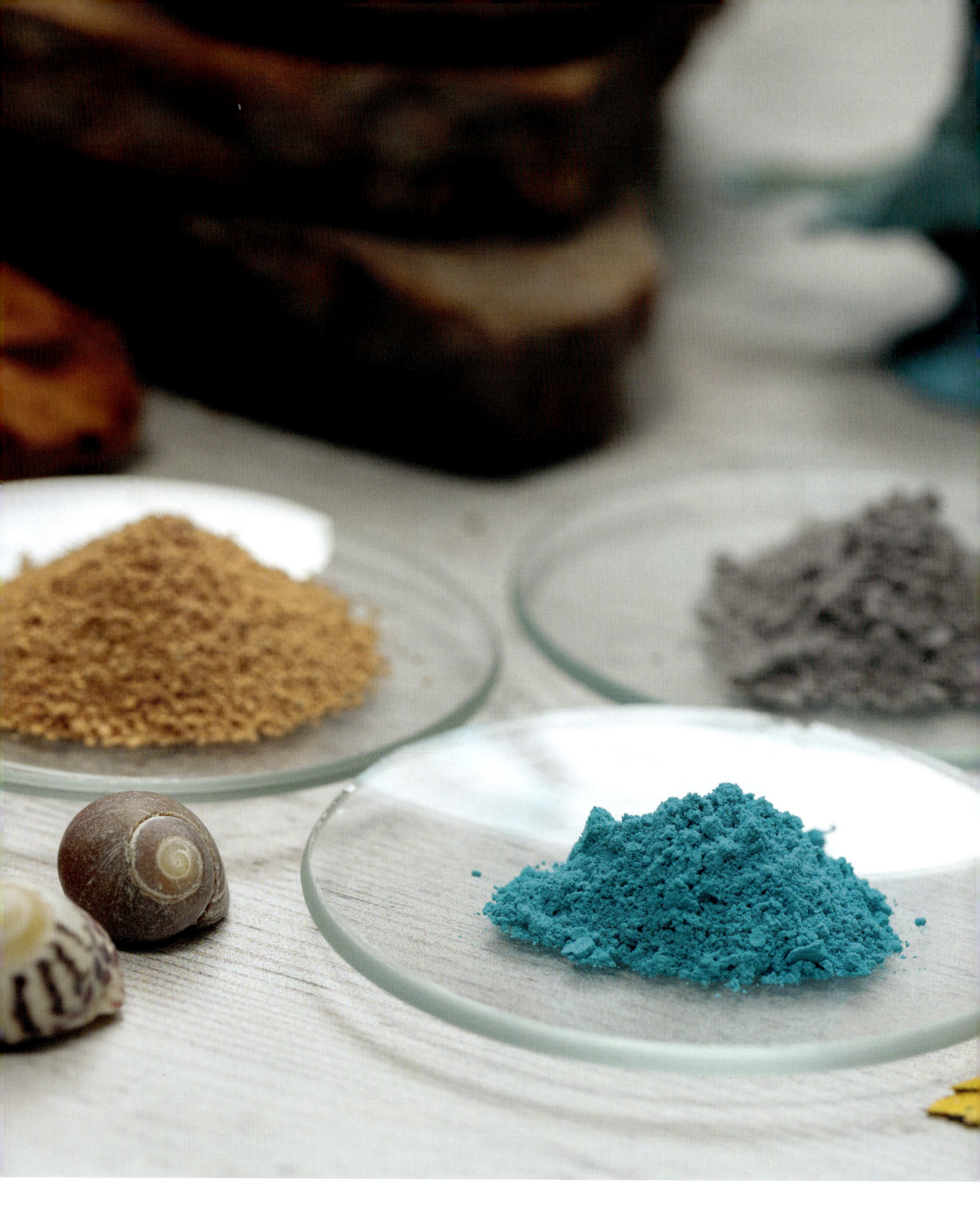

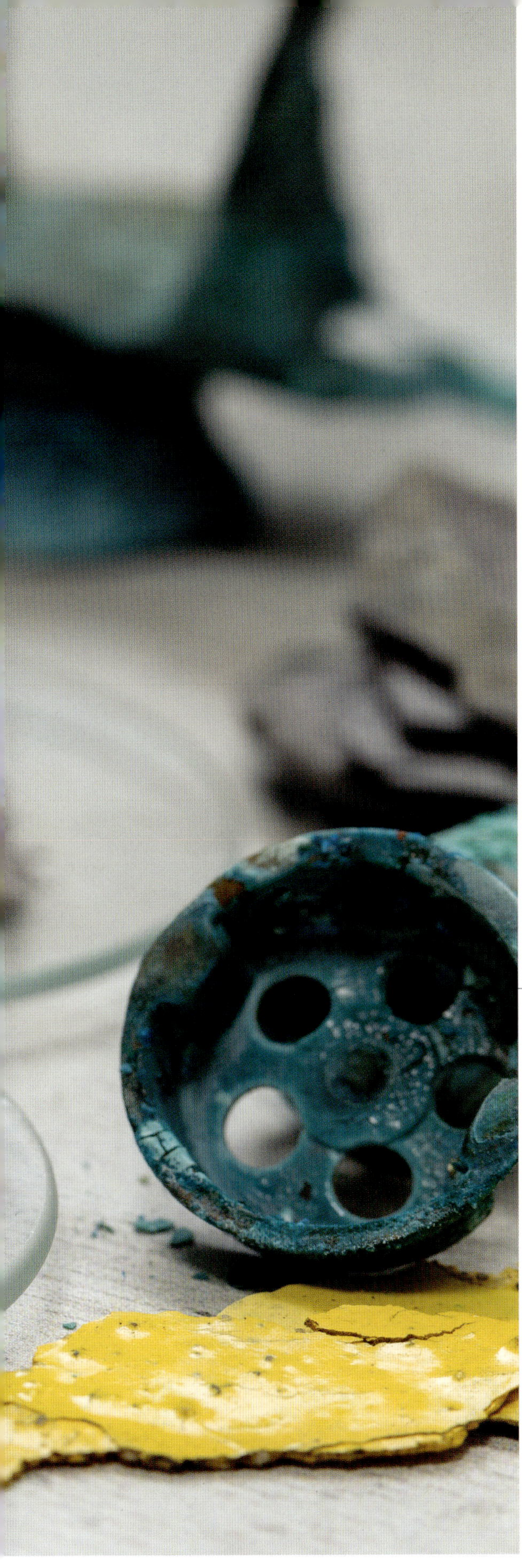

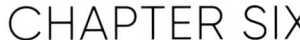

CHAPTER SIX

THE FUTURE OF PIGMENT PRACTICE

As we look to the future, the world of pigments is undergoing exciting transformations. The evolution of pigments goes beyond the confinements of the near past, of widely accepted manufacturing processes or traditional craft techniques of hand-produced colour. There is now a greater permeability between disciplines and bodies of knowledge, combining different ways of thinking about matter with a new ecological awareness.

Pigment manufacturing now extends into the realms of nanotechnology, sustainability, and bioengineering. We are beginning to unpick long-held notions of extractive practices, re-evaluating preconceived understandings of the materials of art making. With increased awareness and questioning, we are no longer dependent on industrially made colours. We are no longer restricted to commercially available colour options.

This chapter focuses on environmentally friendly alternatives to synthetic pigments and the development of new colours using materials procured from waste. With the global demand for safer, more sustainable pigments on the rise, the future of pigment promises to be as colourful as its past. As shown in previous chapters, advancing technology and scientific thinking created highly fabricated and processed pigments that required high levels of expertise, machinery, and accuracy. Today, by re-evaluating the materials that are within our grasp – found within the context of our environments, our homes, and our studios – we can take charge of our consumption and gather local raw materials. We can move pigment manufacture from a global system down to a personal scale, creating a more harmonious relationship with the sources of our pigments.

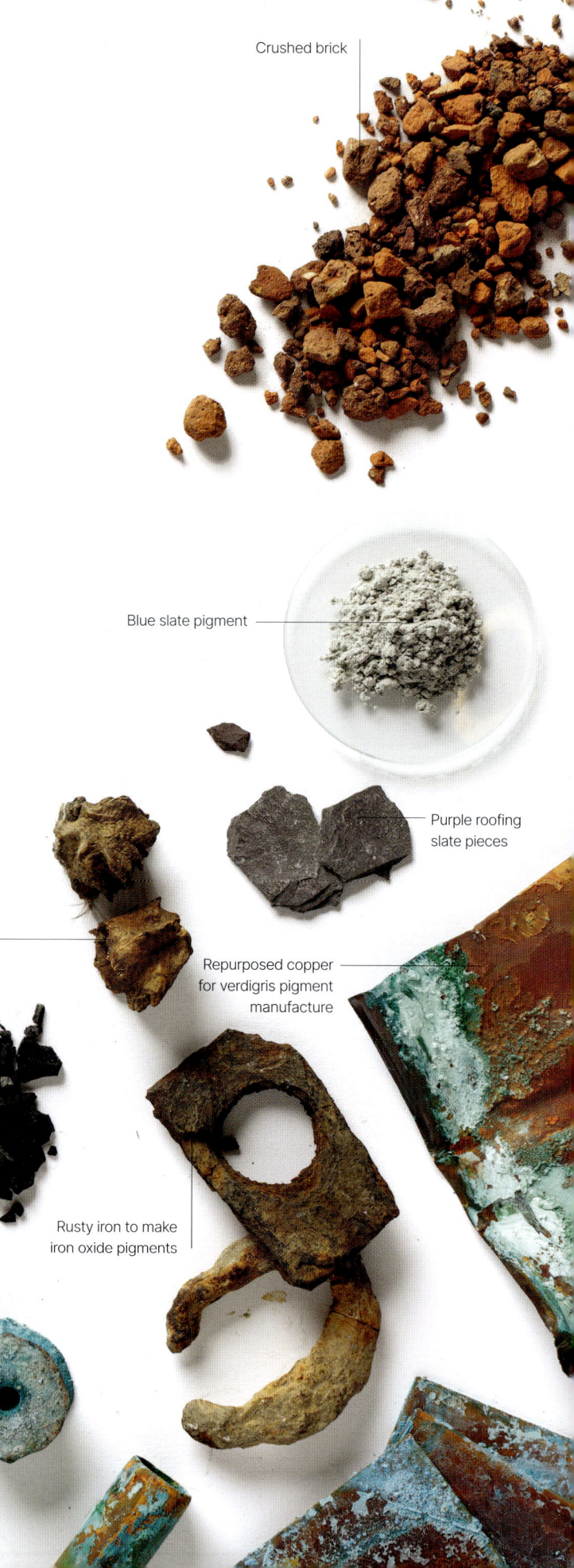

Crushed brick

Blue slate pigment

Purple roofing slate pieces

Repurposed copper for verdigris pigment manufacture

Knopper galls

Georgian yellow brick pieces and processed pigment

Rusty iron to make iron oxide pigments

Found charcoal pieces

AN ARTIST'S ASIDE

RECRAFTING THE MATERIALS OF PLACE

In the summer of 2022, the sky turned a thick murky orange and a soft dusting of Saharan sand landed on the cars and bins on my street in north-east London. Dust storms in the Sahara desert coincided with southerly wind patterns to contribute towards airborne pollution in London. Saharan plumes are produced by greenhouse gasses, especially the over-production of methane.

I made some watercolour paint from this beige sand and filled a tiny Thames shell with it. For me this phenomenon, fuelled by climate change, sits adjacent to the global consumerism we are now at the behest of.

We are members of the natural world. Journeying with pigments, and learning to make them, reconstructs different ways that we can be in the natural landscape. Reconsidering our place within the world in such a way creatively reimagines the tangible nature of our lives. We should be working with and for the sources of our colour to gain mutualistic relationships. A lot of this work is about accepting how much we have gained from natural 'resources' on an individual basis, and taking the time to research where our materials come from. Our collective responsibility will bring about greater effect, using art materials as a lens through which to deconstruct our preconceived values.

Within my work with pigments I have revisited my assumptions in order to begin shifting my thinking. The future of pigment is an active one, the centre of which is about breaking down our material value judgements and hierarchies. Accepting the unique qualities of local materials brings us closer to honouring the starting material. In doing this, fewer by-products are created and less waste is generated. A colour's identity isn't just its colour.

I like the idea of an ever-moving river of materials that is constantly being reused, reimagined, repurposed, where nothing is lost and everything has potential. This way, there are infinite possibilities for just one material, and the presupposed material hierarchies are being re-evaluted.

Right: A piece of 1970's brown brick being transmuted into an artist's grade pigment.

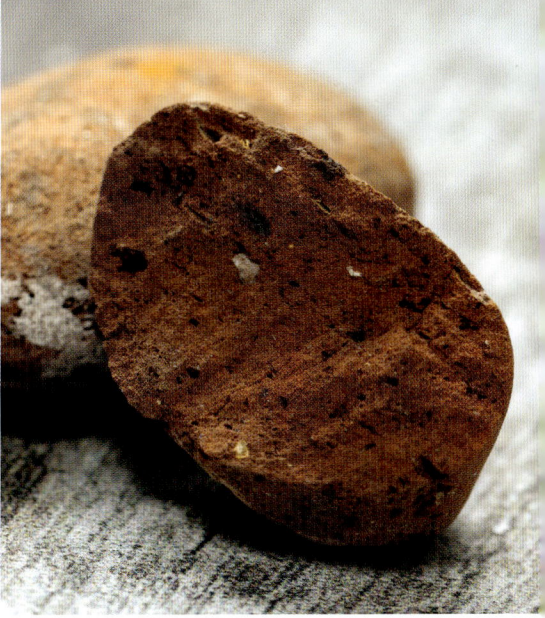

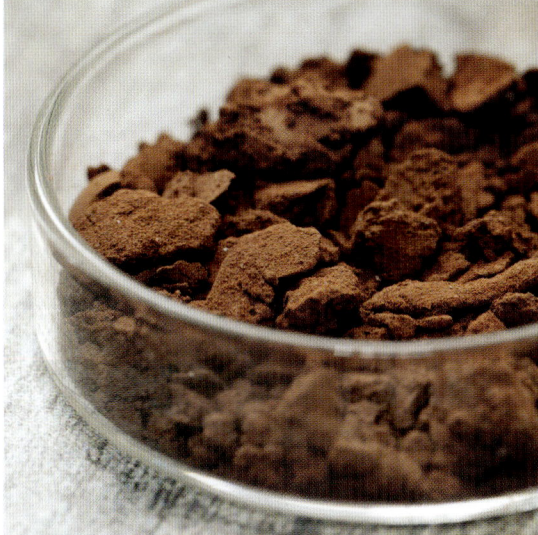

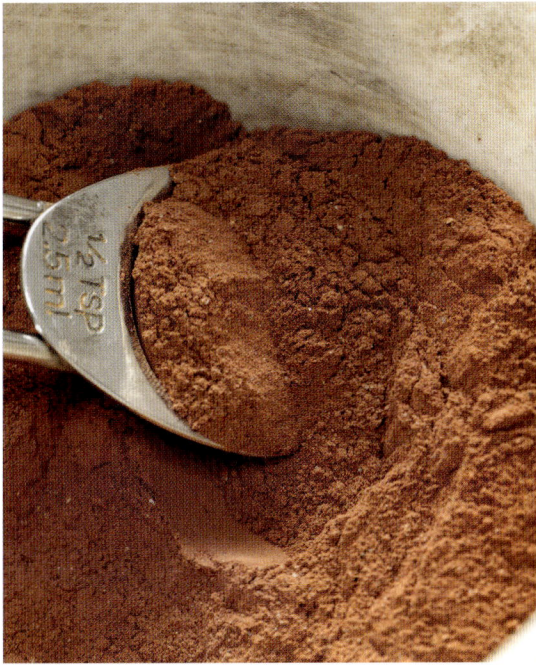

PROCESSING RAW EARTH INTO YOUR OWN PIGMENTS

With the separation from the source of our art materials, as highlighted in Chapter 1: A History of Pigment, it is only fitting that the future of pigment usage be re-evaluated, with a radical reconsideration of the most important group of colours: earth pigments. The incorporation of hand-processed colourants, derived from ethically sourced samples, re-establishes these pigments as the foundation of art-making and beyond.

1. FINDING YOUR SAMPLES

All over the world, there are sources of naturally occurring iron-rich minerals. Look for places where there is little organic soil or decomposing plant material. Good options include places of human intervention (quarries, mines, construction sites), or water-related sites (riverbanks, streams, lakes, seas, dams), or natural features revealing layers of geological time (cliffs, hills, mountains). Let's first consider what you are looking for.

TYPES OF IRON-OXIDE MINERAL SOURCES FOR PIGMENT MAKING

CLAY (ALUMINOSILICATE MINERAL MIXTURES)

Complex material made of a mixture of fine-grained clay minerals, usually less than 2 microns small, formed from the weathering of silicate minerals over long periods of time. The breaking down of these minerals by physical and chemical erosion (such as temperature changes, water, and biological activity) contributes towards a key part of the earth's rock cycle. Clay particles swell when mixed with water to create a malleable and mouldable substance with great plasticity. Types include kaolinite, illite, smectite, chlorite, vermiculite, and sepiolite. They vary in particle size, structure, plasticity, and shrinkage. Clays that contain high quantities of iron and little sand are useful for pigment making.

IRON ORE MINERALS

These usually have an iron content of 50–60% or more, making them incredibly hard and dense; sometimes with a metallic sheen. Generous sources of iron oxide create luminous shades of red and orange when ground finely. Varying in hardness depending on how weathered and oxidized they are, they can be found in lump form within or nearby clay bodies. The best sources of natural ochres are rocks and minerals rich in iron oxides and hydroxides.

HEMATITE (Fe_2O_3)

Found commonly in sedimentary, metamorphic, and igneous rocks, this is the richest source of iron oxide for pigment making, being composed of around 70% iron. It produces luminous red pigments when found either pure or mixed. Iron staining of clay bodies caused by hematite is often referred to as red ochre. It is formed from the oxidation of iron-bearing minerals in a variety of geological settings, including weathered iron-rich rocks and hydrothermal deposits. It is found globally, with substantial deposits in Brazil, Australia, and parts of the US.

GOETHITE (FeO(OH))

A major source of natural yellow ochre, this is a hydrous iron oxide mineral that forms in low-temperature, oxidizing environments. It produces yellow to brown pigments, typically yellow ochre. Goethite forms through the weathering of iron-rich minerals and is often found in soils, bogs, and gossan formations (oxidized portions of ore deposits). It is found worldwide, with significant deposits in France, Spain, and the US.

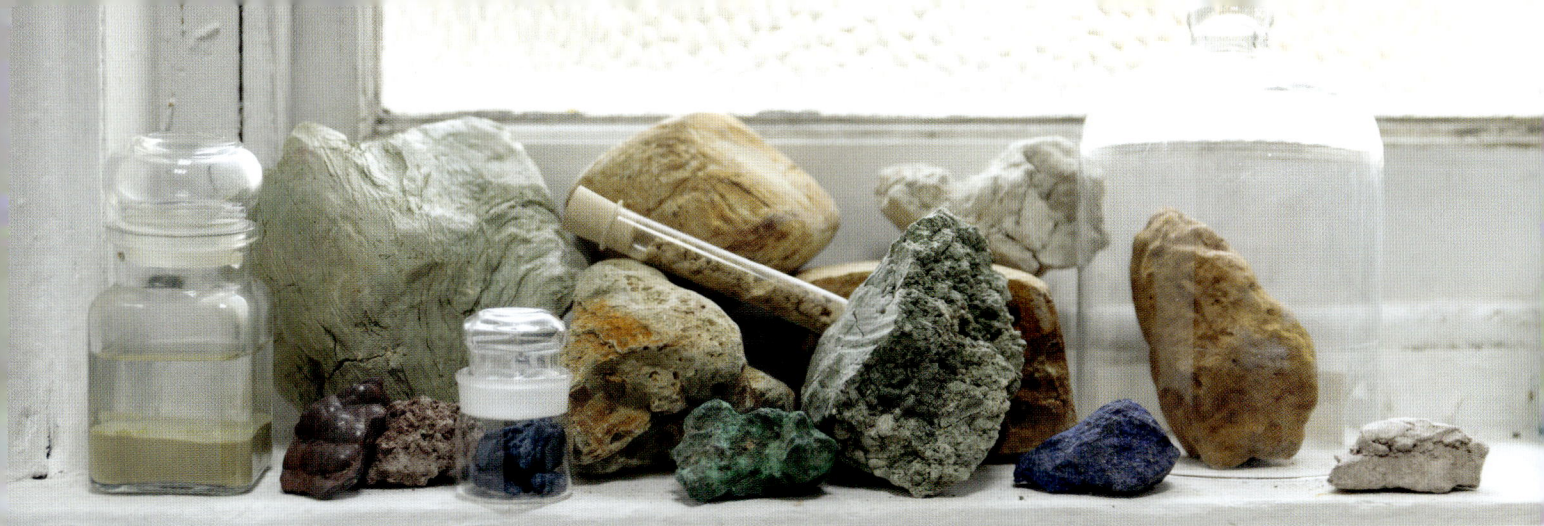

LIMONITE (FeO(OH)·nH₂O)

A mixture of hydrous iron oxides, this is another source of yellow and brown ochres. It forms in weathering environments, often alongside goethite. It ranges from yellow to brown, providing earthy tones. Limonite is typically a weathering product of iron-rich minerals such as pyrite or siderite. Goethite has been referred to historically as limonite and these two names continue to be used interchangeably. Major deposits are found in Italy, France, and the US.

MAGNETITE (Fe₃O₄)

An iron oxide mineral that, when oxidized, can produce pigments, and is more often associated with black or dark-coloured materials. Although it produces darker shades, with oxidation it can lead to pigments similar to those from hematite. It forms in igneous and metamorphic rocks, and is found in beach sands or as a residual mineral from weathering. Magnetite is found globally, with major deposits in Australia, South Africa, and the US.

PYRITE

Pyrite (FeS₂) can indirectly serve as a source of iron oxide through the process of weathering or oxidation. It is an iron sulfide, but when exposed to oxygen and water over time it undergoes chemical weathering to produce iron oxides, such as hematite and goethite, along with other by-products including sulfuric acid. When exposed to oxygen and moisture, it reacts to form ferric iron and sulfate. This ferric iron can precipitate as iron oxide minerals, e.g., goethite or hematite. Iron oxides contribute to the reddish or yellowish colour often observed in weathered rocks and soils. This process is common in gossans, the rust-coloured caps formed over sulfide ore bodies. Pyrite oxidation is a major process in the formation of acid mine drainage, where the sulfuric acid produced can cause environmental damage. However, it also leads to the deposition of iron oxides, in riverbeds and soils for example, contributing to ochre formations.

FERROUS SANDS AND OCHREOUS SEDIMENTS

These sands are typically found along coastal areas or riverbeds and can be rich in iron minerals such as magnetite, hematite, or ilmenite (an iron titanium oxide mineral). Iron staining can also occur in sandstone formations and the red or yellow oxides can be washed from the sand by levigation.

MANGANESE OXIDE

Manganese oxide and iron oxide often occur together in nature, with manganese having an affinity with iron. This combination can produce rich earth pigments like deep umber and sienna, depending on their specific formation and chemical composition. Geological environments that host significant mineral deposits include sedimentary formations (particularly banded iron formations), hydrothermal deposits, lateritic soils, and volcanic regions. Deposits with manganese-iron oxide nodules can create deep brown pigments.

RESEARCH TIPS

You may have little or no knowledge of a site's geology, history, ecology, and land ownership. Before removing any material, here are some points to consider:

- Talk to local people, especially those who know the land intimately.

- Acknowledge Indigenous peoples and seek their permission.

- Identify parties who have a long-standing relationship with the site and ask them to accompany you. These could be individuals or organizations that relate to the local history, ecology or geology of the site.

- Experience the place first without sample removal.

- Use local libraries or town council record offices to find out more about the socio-politics of the space.

- Look through local geological surveys to pinpoint an abundance of iron or iron ore deposits and any safety issues.

- Check the laws both regionally and nationally to see if you may collect there.

- Seek permission from private landowners – people can be very generous when they know you want to work with local materials creatively and respectfully.

YOUR IMPACT ON THE ENVIRONMENT

Roughly 120 million tonnes of clay are mined globally each year by industry. However, the rising trend of using wild clay within creative practices for ceramists, sculptors, and pigment makers is not accounted for within annual statistics.

The direct experience of removing clay is an action that affects the natural world. Considering the specific clay material and your own personal collecting as an isolated exercise is detrimental to the environment. When taking and profiting creatively from a place, there are many ways in which we can give back. You could consider:

- Removing human-made waste from the site.

- Intervening with some local plant husbandry, e.g., help to spread seed, or mulch and prune for growth.

- Learning about the local ecology and inspiring others who live nearby to continue their own relationship with the natural world.

- Creating awareness about foraging techniques through education and workshops.

- Developing ecology at home by broadening and enriching the ecosystems in your garden.

- Donating to charity/organizations associated with the space.

- Taking children to forest school and introducing them to your practice.

- Supporting disadvantaged people to have access to the space and introducing them to your practice.

- Supporting Indigenous peoples who live there.

- Creating your own symbolic offering by spoken word or physical form, such as scattering non-toxic pigments you have made, to embody your intent.

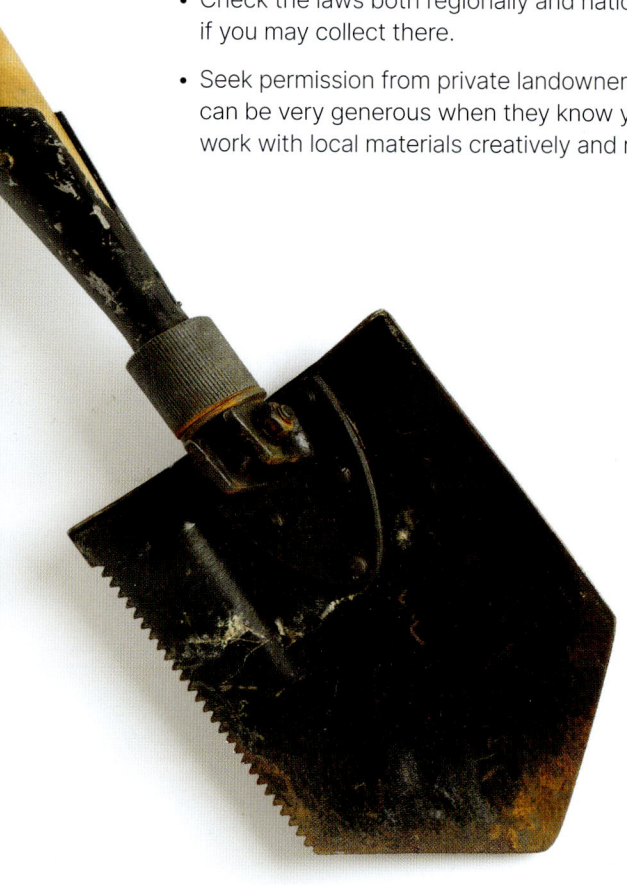

SELECTING YOUR SAMPLE

Once you have considered the ethics of removing material from your chosen site, you are ready to select a rock, a quantity of loose earth or some clay. Choose a starting material that is as soft and crumbly as possible. This limits the amount of physical labour and time required to break it down into a fine powder. Look for the brightest colours – by choosing material with bright, usable pigments, you can avoid taking more than you need.

Colours of different earths in situ can be misleading, often looking brighter against vivid green foliage, or darker if moist. One can get excited about the brightness of a hue only to return to the studio with a bag full of grey dull soil! Carry some sandpaper with you to test the hardness of the material – this is a quick way to gauge the sample colour once ground finely. Or wet the sample with some water and see if any colour can be removed with a brush – if it can, this is a good indication of its potential for pigment.

2. COLLECTING SAMPLES

Once you've assessed the site's safety and the materials present to the best of your ability, put on your protective gear and begin collecting your sample.

EQUIPMENT

- PPE – respirator (P3 rating), gloves and safety goggles

- Trowel (or shovel, spoon or scoop cut from a plastic bottle)

- Bags for samples (waxed cotton, reusable plastic, etc.)

METHOD

Wearing PPE, collect your sample with a trowel or similar. If you locate fallen rocks or disturbed material, use these as this will have a lower impact on the site. If you can do some pre-processing on site – removing small stones, leaves, twigs, etc. – then do so, as this creates less to travel with and ensures less impact on the location. Place the samples in reusable bags and note the GPS location, date of collection, and description of the place.

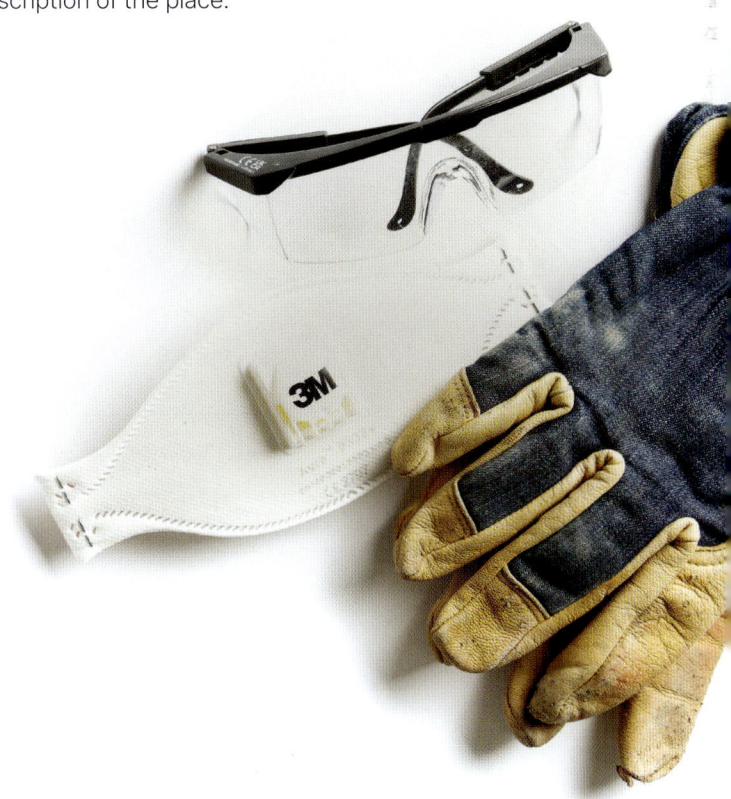

SAFETY NOTES

Rocks that contain silica create incredibly fine airborne particles – once inhaled they do not leave the lungs and contribute towards silicosis, a degenerate and cumulative disease.

Be mindful of heavy metals and radioactive minerals. If there has been some history of mining on the land, there may be areas of contamination or spoil heaps, or uneven ground likely to collapse. Bring a Geiger counter to check for radioactive contamination.

If you're collecting by a water source, locate any safety measures in place and take note of tidal movements.

If material is loose and dusty, wear protective equipment, such as gloves and a respirator.

3. SORTING YOUR SAMPLES

REMOVING UNWANTED MATERIAL

The following methods will remove inorganic material, such as large stones or aggregate; organic material, such as dried or fresh twigs, leaves, and roots; and living organisms (any insects in your sample should be placed outside in a favourable location). If sourcing colour from found material such as a brick, then dust off any mud or dirt using a coarse brush, as organic impurities will contribute towards an unstable paint film and could lead ultimately to the cracking of a painting's surface.

EQUIPMENT

- PPE – respirator (P3 rating), gloves and safety goggles
- Garden sieve or riddle
- Receptacles (bucket, paint kettle)
- Kitchen sieve

METHOD FOR A DRY SAMPLE

Wearing PPE, remove unwanted material either by hand or by using a coarse garden sieve or a riddle with 10mm (⅜in) square holes. Work over a receptacle at least 5cm/2in larger in diameter, to catch material when agitating the riddle. You may need to break the sample up with your hands if it has clumped together. Put stones and plant material caught in the riddle in a garden or take back to the site.

METHOD FOR WET SOIL OR CLAY SAMPLE

Put the sample into a bucket or similar receptacle and add water until a slurry is produced. Some plant material may float to the surface – this can be removed by hand or scooped up with a kitchen sieve. Now it can be put through a riddle as described in the method for a dry sample.

CRUSHING

If the sample is dry and hard, the particle size needs to be reduced, starting with crushing, to prepare it for washing.

EQUIPMENT

- PPE – respirator (P3 rating), gloves and safety goggles
- Strong hessian sack or thick plastic or canvas bags
- Metal mallet or hammer
- Receptacles (bucket, paint kettle)

METHOD

Wearing PPE, put the material in a strong hessian sack or in a thick plastic or canvas bag (double bag the sample), and hit it with a mallet or hammer until you are left with pieces that are less than pea size (5mm/¼in diameter).

GRADING

After crushing, it is possible to remove some pieces that are darker or have inclusions in them. Separate any differences in colour by dividing up the batch, picking out the various pieces by hand. For example, Georgian yellow brick has rather large dark masses in the brick that I try to remove as they make the resulting pigment duller.

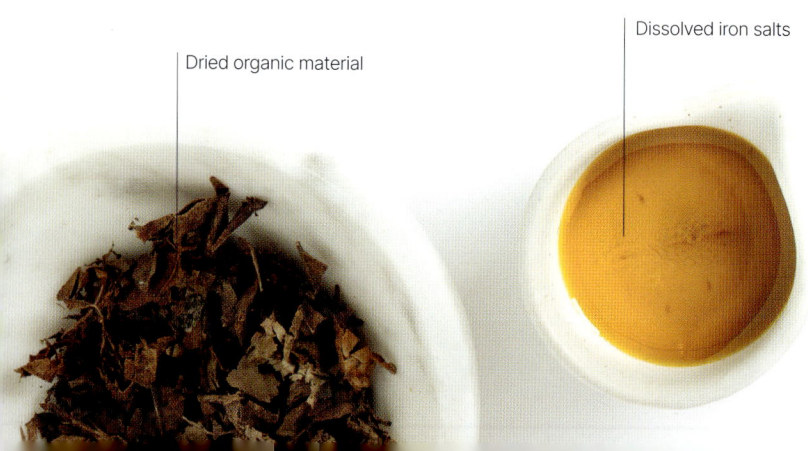

Dried organic material

Dissolved iron salts

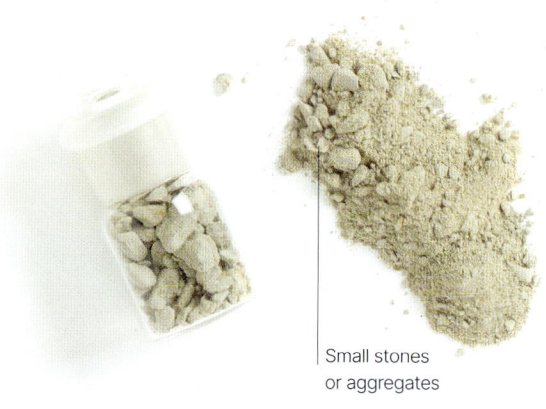

Small stones or aggregates

GRINDING

By now the material is a mixture of fine and coarse dust, and it needs to be ground to be finer.

EQUIPMENT

- PPE – respirator (P3 rating), gloves and safety goggles
- Pestle and mortar
- Electric coffee grinder
- Hand-driven grain grinder
- Ball mill

METHOD

Wearing PPE, grind the material to a powder using a pestle and mortar or an electric coffee grinder. A hand-driven grain grinder can also be utilized.

Depending on the material, using the grinder to break down lumps can take a few minutes or much longer. A ball mill can be used for large batches of material, particularly for hard rubble.

WASHING

The material (such as clay or ochre) is mixed with water to form a slurry.

EQUIPMENT

- Receptacles (bucket, paint kettle)
- Distilled water
- pH indicator strips

METHOD

1. Wearing PPE, pour the dust carefully into large plastic containers and, using gloved hands or a wooden stirrer, mix it with lots of tap water.

2. Wait for the particles to settle to the bottom of the container and pour away the excess water. Soluble impurities such as soluble metal salts, tannins, saponins, and organic resins, and tars such as asphalt can be removed this way. An indication of impurities is a distinctive white, grey, or pale yellow foam in your water and an unclear solution once the particles have settled.

3. Repeat this step until the water that separates is clear. Samples will vary in how much washing they will need – the minimum is usually around five times. The last two washes should be done in rainwater, or filtered water, or, better yet, distilled water.

4. Test the final wash water poured off for its pH. If it is not neutral, continue with the washing steps.

TIP

If you suspect that there are soluble forms of iron in the water jettisoned from your washed sample, it might be possible to convert it into pigment. Iron can be precipitated out of suspension by altering the pH of the solution. This is done by adding an alkali; lime or chalk would work well.

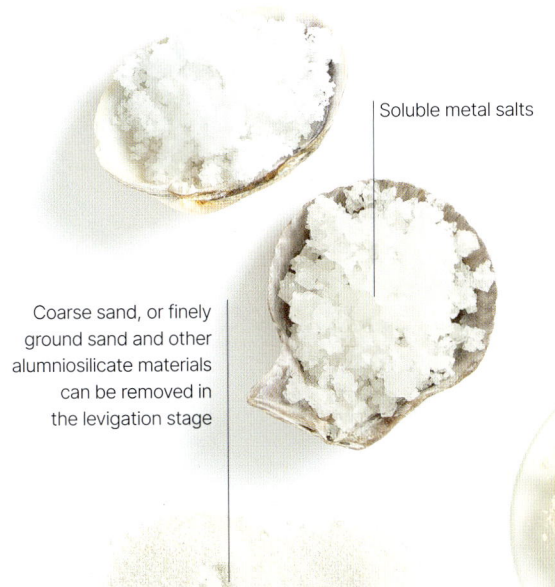

Soluble metal salts

Coarse sand, or finely ground sand and other alumniosilicate materials can be removed in the levigation stage

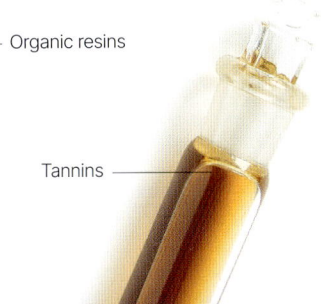

Organic resins

Tannins

4. LEVIGATION PROCESSES

In the production of natural pigments, levigation is used to separate fine pigment particles from coarser impurities. This is particularly important in the preparation of pigments such as ochre, where a uniform texture and consistency are essential for use in paints and dyes. The process involves suspending the material in water and allowing heavier, unwanted particles (such as sand or grit) to settle out, while the finer particles remain suspended. The term is derived from the Latin word *levigare*, meaning 'to make smooth' or 'to polish'.

Levigation creates different grades of the sample, often resulting in differing colour shades. This is a result of the varying grain sizes of the pigment. It is commonly used in industries dealing with ceramics, pottery, pigments, and even cosmetics, where fine, homogeneous particles are desired for better texture and consistency.

EQUIPMENT

- Receptacles (bucket, paint kettle)
- Shallow trays
- Dehydrator
- Pestle and mortar
- Sieve (60 mesh)
- Bottles, for storing

METHOD 1

1. Take a washed mineral sample, and allow it to dry fully. (This enables it to absorb the water added in step 2 very quickly to form a homogeneous slurry.)

2. Make sure all clumps of material are broken up, and fully suspend the finer particles in the water to form the slurry. Agitate the solution so that the fine particles are in suspension.

3. Allow the suspension to stand undisturbed for a few minutes. Heavier and coarser particles, such as grit or sand, will settle at the bottom of the container.

4. Carefully tip the vessel and pour the water containing the finest particles into a second container.

5. Add more water to the first container with the heavier particles in and repeat the pour.

6. The process described in steps 4 and 5 can be repeated multiple times to further refine the material. The resulting samples will range in particle size from sand, to silt, to pigment. The difference in particle size will often create a difference in colour – this is due to the relative masses of the minerals allowing for their separation. The finest sample may be the lightest shade as it contains finely ground alumina silicate material, which is light in colour.

7. Pour the refined, suspended particles into shallow trays and allow to dry, either by natural evaporation or through artificial means in a dehydrator. The samples, upon drying, may clump back together, especially if they contain a lot of clay minerals that have adhesive properties.

8. Carefully pour or decant the material into a pestle and mortar so that it contains a quarter of its volume. Crush and grind the material into a fine powder.

9. Use a fine geology test sieve (60 mesh) to sieve out any remaining lumps. These lumps can be re-ground or disposed of, returning them to the earth in a garden or outside space.

METHOD 2

1. After washing the samples is completed, pour off all the remaining water and leave the remainder to dry.

2. As the pigment dries, the finer pigment will come away from the bottom, leaving heavier pigment as a distinct layer. These layers can be carefully separated by hand. This works especially well if the coarser level is partially sandy and the finer top layer contains finer, greasier clays, as they become very distinct. Continue to process as steps 8 and 9 of Method 1.

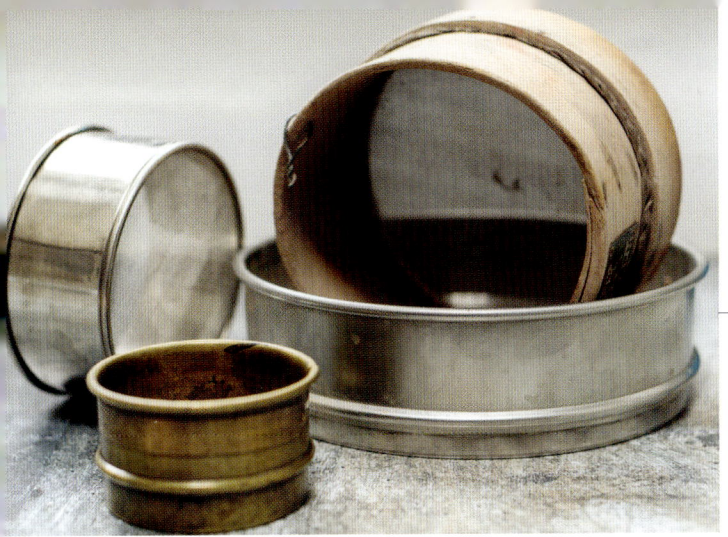

ENGAGING WITH WASTE MATERIALS

We are in the age of the anthropocene. In the last 200 years, humans have arguably made a bigger mark upon the earth than at any other time within human history. Single-use products that cannot be repaired have been produced through industrialized and mechanized processes, as well as vast quantities of by-products and waste.

SIEVE SIZES

Pigment makers use sieves with different mesh and micron sizes to separate particles of various sizes.

The mesh size is defined by the number of openings per inch in the sieve. For example, a 100-mesh sieve has 100 holes per inch, meaning the particles that pass through are smaller than the openings. A larger mesh number indicates smaller holes and thus finer particles, while a lower mesh number indicates larger holes and coarser particles.

The micron size refers to the size of individual particles that can pass through the sieve. One micron (μm) is equal to one thousandth of a millimetre (0.001mm). For instance, a 100-mesh sieve typically allows particles smaller than 150 microns to pass through.

In terms of common mesh and micron sizes, a 20-mesh sieve (850 microns) is used for very coarse material, such as gravel; a 50-mesh sieve (300 microns) for medium-sized particles; and a 100-mesh sieve (150 microns) for very fine pigment particles.

The complete flow or movement of waste – from its point of generation through to its final disposal or recovery – is known as a 'waste stream'. Waste materials can be solid, liquid, or gas, and stages of the waste stream can include collection, transportation, treatment, and disposal. In a broader sense, a waste stream can also refer to a specific category (such as hazardous, organic, recyclable, or electronic), requiring unique management and disposal methods. The term is commonly used in industries and waste management services to track how waste is processed to reduce environmental impact and increase resource recovery.

Traditional and Indigenous cultures historically practise circular systems, where every resource is highly valued, minimizing waste and living in balance with nature. When the use of a material has ended, it can be returned safely to the earth. In Western consumer cultures, waste does not need to be an inevitable feature of capitalism. There are many ways in which we can live with rubbish beautifully. One of these is to reuse and repurpose jettisoned material for use in pigment making.

For those living in towns and cities, access to spaces where traditional pigment materials are bountiful is not easy. Using materials that have had previous use and untold hours of human labour invested in them gives waste-stream pigments exciting and complex meaning. These raw materials are often rooted in the past, but the narrative in which they were collected positions them in the present.

The practice of generating colour from discarded materials to create sustainable pigments reflects the history of a place, and offers an insight into a person's relationship with their surroundings. Colours made from these materials are hues that otherwise would never have become surface colourants. A good example is that of repurposed bricks. They make stable, lightfast colours. Especially when washed, they will be relatively inert within a paint layer. The creation of such colours has a direct link to the calcined earth pigments from Renaissance Italy. Colours such as burnt sienna and umber have long been used as useful deep browns, with sienna leaning towards an orange hue and umber a greenish one. What if we use a discarded red brick from a 1970s building as a pigment in the underpainting of an oil instead of an Italian earth?

A powerful example of repurposing to create art was that of the 'Berlin Butterflies'. These were made by German citizens and Allied soldiers working together in the aftermath of the Second World War. Berlin rubble created by Allied bombing, such as broken roofing, tiles, and brick, was utilized as raw materials for painting. Crushed and mixed with glue, these materials were used to create images of butterflies and birds that were sold as wartime souvenirs, the proceeds of which went a little way towards building a future.

The pigments I make reflect the habitual walks I go on, the places I frequently go to in the world. They are mementos of my own navigation. These materials are mostly waste products from the creation, renovation or destruction of buildings. Within the canon of art, reappropriation of imagery is central to the creation of new work. But why not reappropriate physical material, too, in order to depict ideas and concepts? Waste-stream pigments are sampling from the streets, as opposed to sampling material from past epochs and art movements.

'OVERLOOKED' SOURCES

Sources of raw materials for pigment making might come from commercial or domestic building sites; reclamation yards or road excavations; accessible public places or private property, with permission.

DISTURBED EARTH

Where interventions have taken place to move, retrieve or reveal the geological layers of the earth, deposits may be ripe for removal. When boreholes are dug for concrete foundations for high-rise buildings, for example, the revealed material may be lost or sent to landfill. Clay from deep boreholes loses its original context and is on the way to becoming an undervalued asset, when it might be useful for processing earth pigments.

CONSTRUCTION INDUSTRY DEBRIS

- Broken red and yellow bricks; purple and blue roofing slate

- Garden, driveway, and cement aggregate

- Excavated clay, chalk, and sand

- Lime mortar; plaster

- Copper pipes and lead flashing

- Hard wood; charcoal and coal; wood ash

PUBLIC OR COMMON PLACES

- Public streets, gardens, parks, woodlands (tanin-rich sources: walnut, oak gall, and oak heartwood)

- Riverbanks, ponds, beaches

- Unclaimed land or 'wastelands'

- Unofficial/official dumps, skips, recycling centres

PRIVATE PROPERTY/LAND

- Land estates, farms and smallholdings, garden centres, private gardens

- Food waste: supermarket, restaurant, domestic kitchen

MAKING WASTE-STREAM PIGMENTS

Is it possible to let go of some of the technologies that no longer serve us? Learning about how past civilizations used and made art materials, as well as reusing non-traditional materials, has allowed me to think about pigments differently.

The industrial-scale production of synthetic metal pigments, such as cadmium and cobalt colours, creates by-products that need to be managed accordingly. As with any product with multiple ingredients, it is very difficult to ascertain the traceability of these substances. If we cannot be sure of the impact these products have on their local and global environments, it might be best to avoid them altogether. By abstaining we can reduce the toxic by-products from pigment manufacturing, as well as some dubious ethics of mining procedures, such as the use of child labour in mica mines. If you stick to earth pigments and biodegradable lake pigments and dyes, these can be put back to earth.

Colour isn't solely an abstracted optical phenomenon – it is formed out of the world we live in. Regardless of their starting materials, pigments are made from the earth. Such compassionate reworkings of local materials foster co-existence and empathy with our surroundings. The future of sustainable pigment is not solely dependent on re-evaluating pigments and manufacturing processes – it must also relate to the binders we use. There are limited benefits to reimagining pigments into sustainable substances if we do not consider the vehicles that are used to suspend them. Binders, and by association the effects of paint additives and cleaning agents used in tandem, should also be considered.

I am a collaborator and corroborator with the species of animals and plants from which I garner colouring compounds. I live a fuller and richer life having created a practice that interacts with my surroundings in a deeper and more meaningful way. Take my oak gall ink recipe, for example. By collecting the galls, I have created an intimate relationship with the materials I use – exponentially reaching out from the gall, to the gall creator, to the plant host, to the local ecology and on and on. This relationship is a heady mix of sensory, cerebral exchanges and ultimately is spiritual in nature. Each decoction smells different and feels different. Each ink looks different; every gall creates a different colour due to its unique nature.

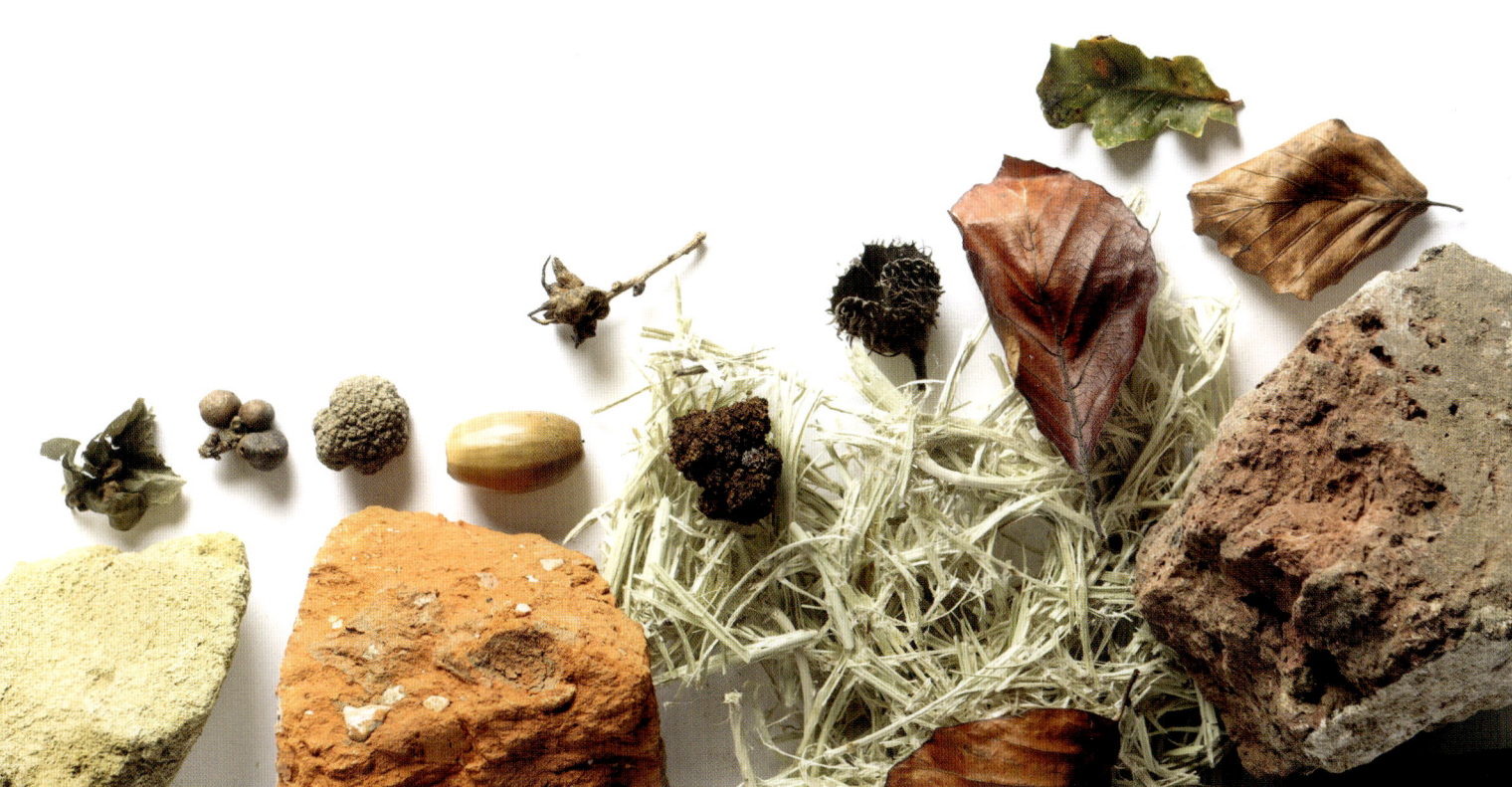

RECIPE: OAK GALL INK

Galls are growths that occur on plants, created when the plant is stimulated by an insect due to feeding or egg-laying. Oak galls are created almost exclusively by delicate miniature wasps. This beautiful and somewhat unusual relationship is what gives oak gall ink its tannin-rich, velvety black finish.

Oak gall ink has been used for some of history's most important documents, including the Domesday Book and Magna Carta (UK) and the Constitution (USA). Despite its potential to corrode the writing surface over time, the ink's permanence made it a favoured medium for scribes, artists, and composers. Its use persisted into the modern era before being replaced by more stable synthetic inks.

Inks made from fallen oak galls destined for a local recycling scheme were one of the first pigments I made from what some might consider 'waste'. I now make many different oak gall inks – the recipe below is what I would call my 'standard' recipe.

INGREDIENTS

- Knopper or English marble galls, 10g
- Distilled water, 100ml
- Iron sulfate, 1g
- Gum arabic crystals, 20g
- Cloves, 50g, or a few drops of clove oil

EQUIPMENT

- Pestle and mortar
- 200ml glass beakers
- Muslin cloth and kitchen sieve
- Funnel
- Inert stirrer (plastic, glass stainless steel)
- PPE – gloves

METHOD

PREPARING THE OAK GALLS

1. Crush oak galls using a pestle and mortar into roughly pea-sized pieces. Soak galls in distilled water for one week to ferment.

2. After soaking is complete, simmer galls and cloves in distilled water for 30 minutes and let cool. Leave for one week, stirring daily.

MAKING THE INK

1. Strain the oak gall mixture (decoction) through a piece of muslin cloth in a kitchen sieve, then add the iron sulfate.

2. Crush gum arabic crystals in a pestle and mortar. Add 50ml of boiling water and stir until the crystals are fully dissolved.

3. Add the gum arabic solution to the gall decoction and decant into a bottle using a funnel.

NOTE

When collecting specimens of oak galls, try to do so mindfully. If you suspect your specimens still contain wasp larvae, make sure that when they appear from the gall they can be safely released. Collect galls from where there are plentiful supplies, and avoid returning to the same place over consecutive years. Organizations such as The British Entomological and Natural History Society, or the British Plant Gall Society have useful guidelines on oak gall collection.

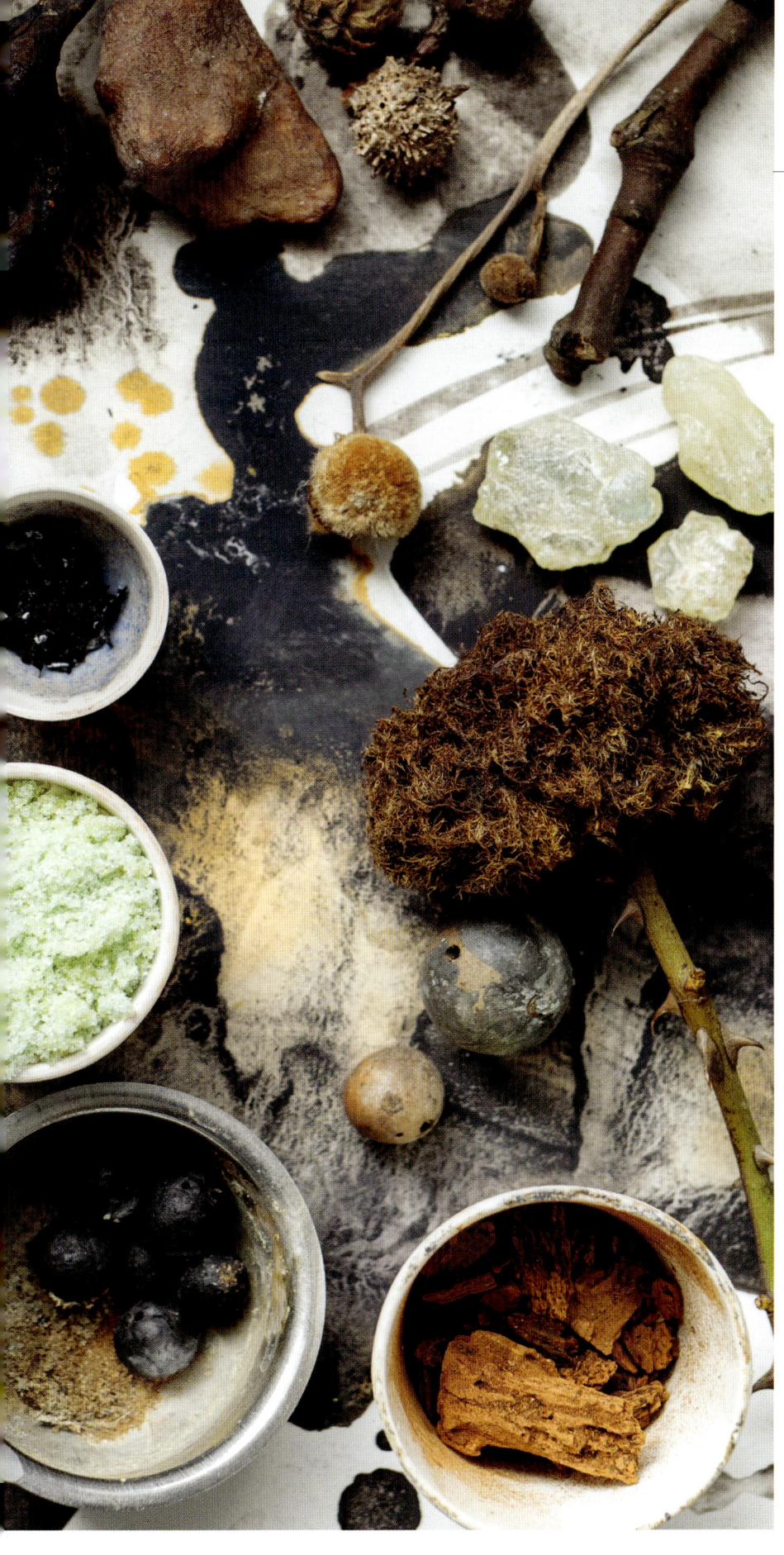

PIGMENT SUMMARY

Pigment Name: Iron gall black gallotannin

Nomenclature: Iron gall, oak gall, iron-tannate, gallo-tannate

Pigment Index Number: Not applicable

Colours: Black, brown-black, blue-black

Chemical Name: Gallotannin

Chemical Formula: $(Fe_3(C_7H_5O_6)_6)$

Classification: Synthetic organic

Description: Gallotannins, a class of hydrolysable tannins derived from gallic acid and glucose, are significant in the production of natural pigments and dyes

Conservation Issues: Can degrade the substrate over time due to acidic ingredients

Particle Morphology: Irregular, amorphous

Dates in Use: Roman era to 19th century

Density: 1.2 to 1.4g/cm³

Hardness: 3 on Mohs scale

Oil Absorption Ratio: 30–50g of oil per 100g of pigment

Refractive Index: 1.50–1.60

Health and Safety: Considered non-toxic in small amounts used, but take usual precautions when handling fine dusts

Environmental Impact: Gallotannin pigments derived from plant-based sources are relatively eco-friendly if sustainably sourced. Be mindful of the ethical collection of gall species. Iron salts used in gallotannin production are acidic and, while commonly used for soil health, should be disposed of carefully to prevent environmental harm.

WORKING WITH VERDIGRIS AND COPPER

Verdigris is a term used to describe a mixture of copper salts – primarily copper acetates and basic copper carbonates – that form on copper and copper alloys, such as bronze surfaces exposed to the environment over time. The process starts with oxidation. Copper reacts with oxygen in the air to form copper oxide. The copper oxide then reacts with carbon dioxide and water from the atmosphere to form basic copper carbonates, a form of synthetic malachite. In the presence of acetic acid (from sources such as pollution, or perhaps very minimally from bacterial action in specific environments), copper acetate (verdigris) can form.

The historical recipes for creating verdigris pigments show a fascinating variety in copper sources and forms, yet reproducing these recipes is notoriously challenging. The old instructions often feature vague measurements, and the raw materials and environmental conditions have changed significantly since then. Today, we sometimes rely on store-bought or laboratory-grade ingredients, which lack the impurities that would have been present historically. For instance, fermenting apple skins to create organic vinegar, naturally mixed with dust and sediment, yields a vastly different pigment from a lab-prepared 10% acetic acid solution with distilled water, even if both have similar acidity levels. These 'impurities' in handmade reagents allow for a rich spectrum of unique hues.

Create your own variants for the following verdigris recipes by experimenting with copper and acids sourced from waste streams that are local to you. Experimenting with found materials – old plumbing pipes, discarded brass hardware, second-hand shop trinkets – can amplify the wonder of these colour transformations. Street finds, riverbanks, seashores, and even the odds and ends in a drawer or garden shed can offer intriguing sources for verdigris-making.

While artists have prized certain pigments in the past for their colour, rarity, symbolism, and stability, verdigris pigments are highly transparent and reactive, and the colours are unpredictable. Copper-containing compounds can range in colour from clear, bright turquoise blues, to soft mint greens or deeper grassier shades of green. I find that exciting, and I am not alone: hashtag verdigris on most social media platforms for inspiring strange blue concoctions the world over.

Verdigris gets a bad press for being a colour that can darken or go brown with time if not treated accordingly, but this is precisely why I like it. It does what it wants. Its shape-shifting colour (due to moisture, air or the presence of sulfur) is a thrilling ephemeral prospect. It constantly surprises me. Each batch is a little different – the way the crystals form, their size, their colour – it's all so unpredictable. Yet anyone can make it, using any copper (or alloy) and any acid (lactic, acetic, citric), then adding some extra bits, maybe a catalyst (salt), or a filler and neutralizer (chalk), or a few drops of ammonia for a madly intense electric blue.

SAFETY NOTE

Copper acetate pigments are considered toxic to humans and aquatic life, so they should be used with caution. Always wear gloves and a quality respirator or dust mask. Keep away from children and pets, and dispose of safely.

RECIPE: HISTORICAL VERDIGRIS PIGMENTS

As demonstrated by the following recipes, verdigris exhibits remarkable variation in its composition and colour, and for this reason, they are presented without individual pigment summary boxes. The formation of these corrosion products is highly influenced by local humidity and atmospheric conditions, which affect both the rate of formation and the materials produced, and once the underlying scientific principles are understood, there is ample opportunity for experimentation, with emphasis on pigment making as a unique space or locus for a multidisciplinary, intuitive craft. These recipes are inspired by historical traditions from different eras.

RECIPE 1: COPPER CARBONATE HYDROXIDE
ANCIENT EGYPT, 3,000 BCE

Also known as natural verdigris, this form is produced when copper reacts with carbon dioxide and water vapour in the air, forming a greenish-blue patina. This reaction is slower and occurs naturally when copper is left exposed to air and moisture. Chemically similar to naturally forming malachite, it could be described as a synthetic malachite pigment.

INGREDIENTS & EQUIPMENT

- Copper plates

- White vinegar

- Jar or vessel

- Scraper and pestle and mortar

METHOD

1. Expose copper plates to humid air in an enclosed space, sometimes with exposure to mild acids like vinegar. Over time, the copper will react with carbon dioxide in the atmosphere, forming a layer of copper carbonate hydroxide, alongside copper acetate.

2. Scrape the verdigris into a pestle and mortar and grind into a fine powder.

RECIPE 2: COPPER CHLORIDE
1ST CENTURY CE

This form of verdigris is produced when copper is exposed to salt (sodium chloride) and moisture. It was sometimes used for decorative painting and in the colouring of glass and ceramics. Less stable, it was prone to darkening over time.

INGREDIENTS & EQUIPMENT

- Copper plates

- Distilled water and salt

- Jar or vessel

- Scraper and pestle and mortar

METHOD

1. Place copper plates in a container with saltwater (use 20g of salt per 100ml distilled water used). The plates can be partially but not wholly submerged. Alternatively, rub salt onto the copper surface and allow it to react with atmospheric moisture – this can be done in an open jar or laid upon a tray. Leave for the basic copper chloride to form and build up on the copper over time.

2. Scrape the verdigris into a pestle and mortar and grind into a fine powder.

RECIPE 3: COPPER URINATE
5TH–18TH CENTURY CE

Combining urine and hay introduces multiple organic acids and compounds, creating a unique environment that can result in a verdigris with unusual characteristics. This recipe will create a deep teal pigment with a blue undertone.

INGREDIENTS & EQUIPMENT

- Copper plates
- Stale urine
- White vinegar (optional)
- Hay or straw
- Clay pot or barrel with cover
- Distilled water
- Scraper and pestle and mortar

METHOD

1. Stale the urine for several days to increase ammonia content. As urine ages, the urea decomposes into ammonia, which is key for the verdigris formation process. Optionally, for a more intense reaction, you can add a small amount of vinegar to the urine.

2. Line the bottom of a clay pot or barrel with hay or straw. Place clean copper plates, free of oils or dirt, on top and pour the stale urine over them, ensuring partial submersion.

3. Cover and allow the mixture to sit for 1–2 weeks. The hay provides organic acids and additional surfaces for verdigris to form, while the ammonia in the urine accelerates the process.

4. Once the verdigris forms, scrape it off, rinse with distilled water, and dry. It can then be ground into a fine powder using a pestle and mortar.

PIGMENT SUMMARY

Pigment Name: Verdigris

Nomenclature: Copper acetate; copper, Spanish, or salt green

Pigment Index Number: PG 20

Colours: Green to blue-green

Chemical Name: Copper(II) acetate

Chemical Formula: $C_4H_8CuO_4$

Classification: Synthetic inorganic

Description: Forms through the reaction of copper metal with acetic acid in the presence of oxygen, creating a copper acetate compound

Conservation Issues: Prone to instability, especially when exposed to light, humidity, or acidic conditions; may degrade, becoming dull or even brownish, and corrode the underlying paint layer or canvas or wood surface

Particle Morphology: Fine, crystalline, typically irregular, with a smooth texture. Plate- or needle-like in structure; size can vary depending on the manufacturing process

Dates in Use: From ancient civilizations (Egyptians, Greeks) through to the Renaissance and into the 19th century

Density: 40g/cm³ (varies with water content)

Hardness: 2–3 on Mohs scale

Oil Absorption Ratio: 35–45g of oil per 100g of pigment

Refractive Index: 1.50–1.80

Health and Safety: Can be toxic, especially if inhaled or ingested in large quantities; handle with care

Environmental Impact: Harmful to aquatic life when disposed of in huge quantities but small amounts can be diluted with water before household drain disposal

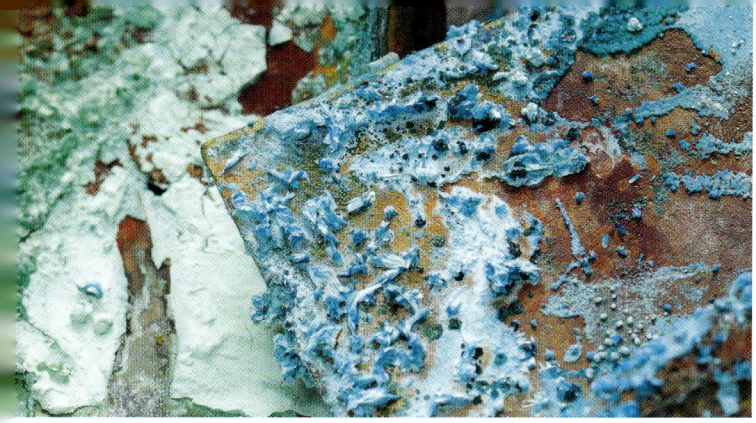
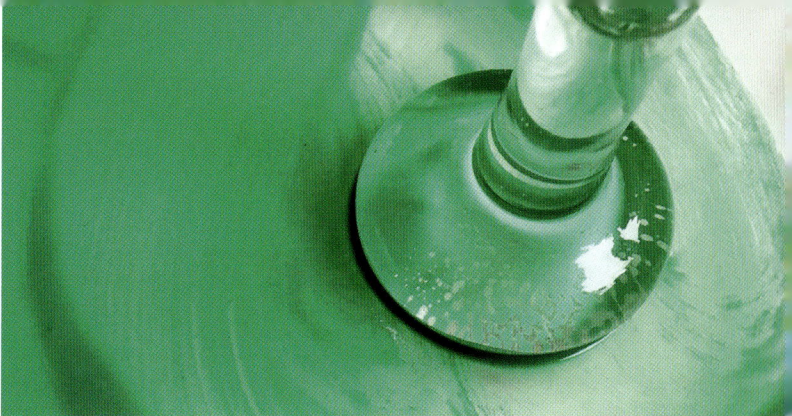

RECIPE 4: BASIC COPPER ACETATE
6TH–9TH CENTURY CE

Basic copper acetate is the most common form of verdigris, produced when copper is exposed to acetic acid (the acid present in vinegar). It forms a greenish-blue crystalline substance as a corrosion product on the surface of the metal. It was widely used as a green pigment in medieval and Renaissance manuscript illumination, frescoes, panel paintings, and for tinting glass.

INGREDIENTS & EQUIPMENT

- Copper strips
- White vinegar
- Jar with lid
- Scraper and pestle and mortar

METHOD

1. Place strips of copper in a jar filled with some vinegar and seal the jar. The copper reacts with the acetic acid vapours in the vinegar and in the air over time, forming a crust of verdigris on the copper surface.

2. Scrape the verdigris into a pestle and mortar and grind into a fine powder.

RECIPE 5: DISTILLED COPPER ACETATE
9TH–12TH CENTURY CE

Also known as neutral copper acetate, this verdigris is formed when copper is exposed to a stronger acetic acid solution with minimal water. It has a bluish tint and is more crystalline than basic copper acetate. Its translucent properties were preferred in oil-painting glazing techniques during the Renaissance.

INGREDIENTS & EQUIPMENT

- Copper plates
- White vinegar (or distilled water and acetic acid)
- Container with lid
- Scraper and pestle and mortar

METHOD

1. Soak copper plates in concentrated vinegar (or with acetic acid and distilled water in a 1:10 ratio) in a sealed container for several weeks. Ensure the copper is submerged to minimize exposure to air, and therefore to impurities. The neutral copper acetate forms as a bluish layer on the surface of the copper.

2. Once the solution is a saturated blue colour it can be evaporated.

3. The verdigiris produced can be mixed with more vinegar and then dried to form a more homogeneous and stable product known as 'distilled verdigris'.

4. Scrape the verdigris into a pestle and mortar and grind into a fine powder.

PROTECTING COPPER PIGMENTS FROM DISCOLOURATION

Copper acetate's solubility plays a key role in its reactivity within paints. Being soluble in solvents such as water and alcohol, copper acetate can dissolve and interact with other components in the paint mixture. This solubility allows copper ions to participate in chemical reactions with binders, other pigments, and environmental factors, causing changes in the paint's colour and stability over time. In oil paints, for example, copper acetate reacts with fatty acids, forming copper salts like copper oleate, which alters the pigment's properties and transparency.

The solubility of copper acetate also makes it more susceptible to oxidation and degradation when exposed to moisture, air or light. This leads to colour shifts and fading, and to the formation of copper carbonates and sulfides. Verdigris pigments also experience these reactions. When exposed to air, copper reacts with carbon dioxide and moisture, creating copper carbonate and causing the pigment to shift from vibrant green to a duller hue. This process is accelerated by humidity. Additionally, verdigris is sensitive to sulfur compounds found in the air, such as pollution, which can further degrade the pigment, turning it brown or black as copper sulfide forms; and extended exposure to UV light can also break it down, altering its colour over time.

When verdigris is mixed with water-based binders like gum arabic, the paint film is weaker, making the pigment more vulnerable to environmental factors. In damp environments, water-soluble paints soften, accelerating colour changes and offering less protection to the pigment. To protect copper-based pigments, they can be encapsulated in protective layers, such as a varnish, a method used by European artists from the 15th to the 17th century to safeguard their works. Verdigris can also be mixed with drying oils, such as linseed oil, where it partially dissolves to create a transparent paint. This technique, which forms copper oleate, results in a luminous effect ideal for glazing layers. Copper oleate, an organometallic compound belonging to the class of metal soaps (copper salts of oleic acid), was historically used in anti-fouling paints, but its toxicity to marine organisms raises environmental and health concerns today.

When mixed with natural resins like damar, copper resinate is created. This pigment is soluble in organic solvents such as turpentine and oil, allowing it to be tempered into oil paint using a muller or palette knife. Renaissance artists favoured copper resinate for its vibrant green colour and transparency, often using it for foliage, drapery, and other elements requiring rich, translucent greens. The process of making copper resinate involved dissolving verdigris in warm resin (often Venice turpentine), creating a green-coloured resin that could be thinned with turpentine or oil to form a glaze.

However, the use of resinate and oleates in paint technology can cause issues such as efflorescence, where crystalline substances appear on the surface of the artwork. The formation of copper oleate can contribute to the deterioration of historical paintings, leading to structural instability in the paint film. This can result in cracking, flaking, or delamination of the paint layers from the substrate. Understanding the formation of copper oleate is crucial for diagnosing and treating the deterioration of artworks.

RECIPE 6: COPPER LACTATE
18TH–19TH CENTURY CE

This Russian recipe, used by icon painters from the Middle Ages to the late 19th century, produces a protein-rich copper pigment made from milk that has gone sour inside a copper pot. Due to the antimicrobial properties of copper, mould growth is unlikely. The pigment's stability is likely enhanced by its protein-laminated particles, with the copper coating playing a key role. For optimal results, allow the milk to react with the copper for three months.

INGREDIENTS & EQUIPMENT

- Copper plates
- Sour milk or buttermilk (use unhomogenized/raw milk)
- White vinegar (optional)
- Copper container with copper lid
- Distilled water
- Scraper and pestle and mortar

METHOD

1. Place copper plates in a copper container with a copper lid. Pour the sour milk over the copper plates, ensuring that they are partially submerged.

2. Keeping it covered, let it sit in a warm area for 1–2 weeks. The lactic acid in the milk will react with the copper, forming verdigris.

3. Once verdigris forms, scrape it off, rinse with distilled water, and dry. It can then be ground into a fine powder using a pestle and mortar.

HISTORICAL AND CONTEMPORARY BINDING MEDIA

In order to adhere the pigment particles to a surface – for example vellum, parchment or cotton papers – a binder is added. In modern watercolours and water-soluble inks, gum arabic from the Sahel region of Africa and Arabia is used. Low grades derive from Nigeria, with the highest grades hailing from Kordofan in Central Sudan. Gum arabic is a naturally occurring dried exudate collected from the trunks and branches of several species of the acacia tree.

Consisting of various polysaccharides (starches) as well as glucuronic and galacturonic acids, gum arabic is a useful adhesive, emulsifier, and thickener. This important commodity was exported from as early as the 1400s to Europe, and in the 1800s became a major export from the British colonies in West Africa to the UK. It was, for hundreds of years, an exotic and expensive material. Saps procured from species of fruit-bearing trees, such as cherry, peach, and plum, were used as alternative local sources for water-soluble adhesives. Common ivy has also been used as a source of gums for watercolour. Gum arabic was used exclusively for commercial industry by the early 19th century.

I make a water-soluble binder from the sap of the fruit-bearing species of the cherry tree, but other species of the *Prunus* genus could also be used. This gum is collected in late summer, cleaned and dissolved in hot water to make a yellow-brown liquid. Cherry sap is a little weaker in strength than gum arabic, so I use a more concentrated solution for my inks and watercolour paints.

RECIPE: LOCAL WATERCOLOUR BINDER

Watersoluble tree saps from species of trees from the *Prunus* genus produce useful gums that are similar to, but not as strong as, gum arabic.

INGREDIENTS

- Tree sap from species of the *Prunus* genus (cherry, apricot, plum)*
- Honey or vegetable glycerin (optional)
- Essential oil such as clove, rosemary or lavender)
- Distilled water

EQUIPMENT

- Pocket knife
- Receptacle
- PPE – gloves
- Funnel and muslin cloth
- Pestle and mortar
- Brush and watercolour paper

I have used fresh sap from species of fruit-bearing cherry. Note: ornamental varieties of cherry do not produce the right sap.

METHOD

GATHERING THE SAP

1. Collect fresh sap in the late summer or early autumn when trees have stored the most and will sometimes have an excess exuding from their trunks. Peel off hardened tears carefully by hand, or gently tease away with a knife, trying not to break the bark unnecessarily. If the sap is soft, use your knife to scrape it into a receptacle.

2. Sap that is very dark in colour will contain more sugar and tannin than lighter coloured samples – reserve these for making paint with darker coloured pigments.

MAKING THE BINDER

1. Dissolve hardened sap by soaking it in boiling water in a 1:1 ratio until it becomes a thick liquid. Pour the hot water over the sap and mix rigorously until dissolved. Filter the solution through a funnel lined with a muslin cloth.

2. Adjust your solution based on consistency. The goal is to achieve a smooth, syrupy mixture that isn't too sticky or thick. Sap from cherry trees is approx. one-third less strong than gum arabic. It is easier to mull less viscous binding mediums, and your mixture will start to evaporate on mixing so more water might be needed.

3. To prevent the binder from becoming brittle when dry, add a few drops of honey or vegetable glycerin. This optional step will help keep your paints flexible and easier to rewet later. Do not add more than 10% in volume or your paint will struggle to dry thoroughly and may remain tacky and attract dust. If storing the binder for a while, add a drop of essential oil as a natural preservative.

4. Stir the mixture until it is consistent and smooth.

USING THE BINDER

1. Combine the binder with powdered pigment in small amounts at a 1:1 ratio, grinding in a pestle and mortar until the paint reaches your desired colour and texture. Once well combined, test with a wet brush; dilute if needed.

2. Test on your chosen substrate (paper, board, etc.); more absorbent surfaces require less binder. Adjust ratios if needed. If too shiny, there is too much binder; if too sticky, there is too much honey or glycerine; if too matte, or the pigment brushes off the paper, there is not enough binder.

3. Store the binder in an airtight container in a cool, dry place out of direct sunlight. Keep it in the fridge to stop mould growth.

THE NUANCES OF PIGMENT MANUFACTURE

In handmade pigment production, a spectrum exists regarding the degree to which raw materials are modified. Lake pigment manufacture, combining natural dyes with laboratory-grade chemicals to create insoluble pigments, is a compelling example of this complexity. For instance, is the use of imported alum, extracted through mining, a sustainable way to create ethical biodegradable organic pigments? Such questions highlight the need to balance authenticity with practicality.

Natural pigments may be valued for their historical and aesthetic significance, while synthetic versions are often more stable, cost-effective, and consistent, important qualities in both display and conservation contexts. Navigating the interplay between these considerations is essential, particularly in the domains of curatorial scholarship and conservation practice. While natural pigments may be preferred for their historical fidelity and aesthetic nuance, synthetic counterparts offer advantages in terms of stability, cost-efficiency, and reproducibility, key considerations in both display and preservation strategies.

RECOGNIZING AND ACKNOWLEDGING

Preserving traditional materials remains essential. The synthesis of historically significant or endangered pigments, such as natural ultramarine or Kermes, not only ensures their continued study and application but also mitigates the depletion of scarce natural sources. These concerns intersect with broader ethical and environmental considerations, positioning pigment research within a dynamic, interdisciplinary, and socially evolving field. There is growing scholarly interest in the relationships between material provenance, cultural identity, ecological responsibility and stewardship. Increasing focus is being placed on the intricate relationships among materials, their places of origin, and the communities with which they are intertwined, both human and the beyond-human.

Finding an equilibrium between authenticity and practicality involves paying close attention to supply chains, environmental impact, and the preservation of traditional knowledge systems. By acknowledging the contributions of traditional knowledge holders and considering the origins of our materials, we can honour the earth through a more intentional and respectful pigment making practice.

Right: Azurite and malacite in their natural form, alongside their synthesised analogues.

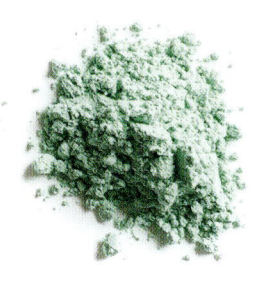
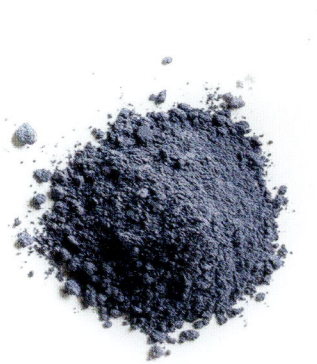
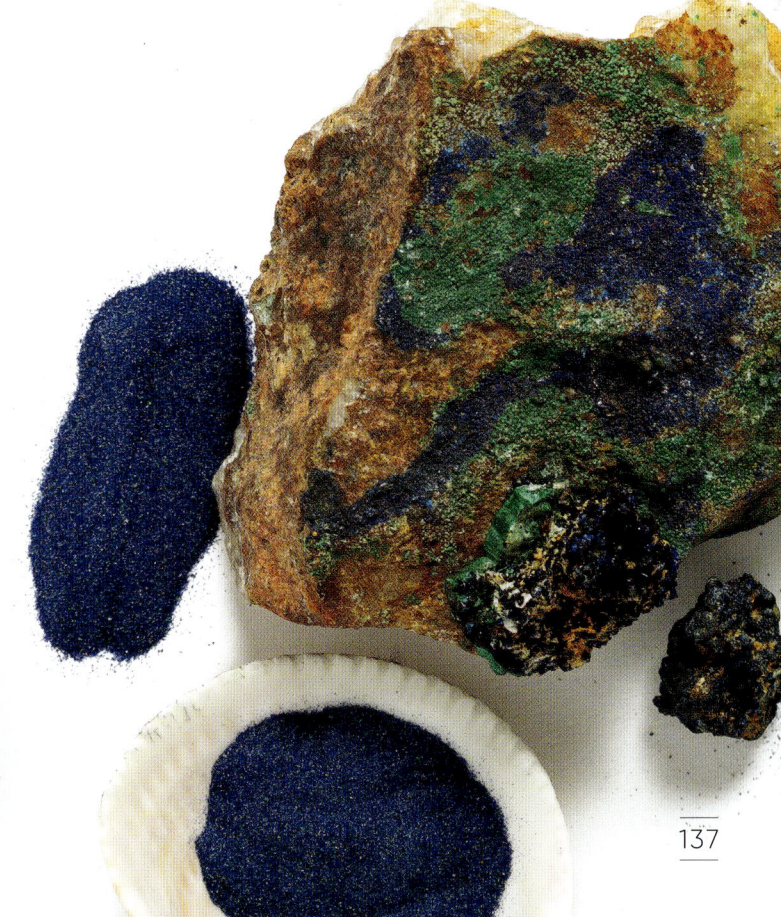

VEGAN ALTERNATIVES

Traditional, non-toxic binders often rely on animal derivatives like gelatin (derived from animal collagen) and casein (from milk). These proteins form strong, durable bonds in paints, varnishes, and adhesives by creating films as they dry, while egg tempera, an animal-based lipid binder, imparts smooth, glossy finishes to paintings due to its emollient properties. However, with growing demand for vegan alternatives driven by ethical concerns, environmental sustainability and dietary preferences, plant-based binders are becoming increasingly popular.

Plant gums and starches, derived from a variety of plant sources, are common vegan binders. Gum arabic, for example, is harvested from the sap of acacia trees in sub-Saharan Africa and is often used in watercolours and inks. Agar, derived from red algae, and guar gum, from the guar plant, are also used for their gelling and binding properties. Starches, such as those from corn, rice or potatoes, are widely utilized for their versatility in adhesives and paints, offering a non-toxic, biodegradable alternative to animal-based glues and binders. Aquafaba from chickpeas is an excellent matte binder and serves as a great substitute for gum arabic.

Soy protein, extracted from soybeans, is a plant-based protein binder commonly used in a variety of applications, including paints and adhesives. Wheat gluten is another plant-derived protein that can be used as a binder in art materials. These plant proteins offer similar properties to animal-based proteins, forming strong, flexible films.

ENDNOTE

The future of pigment may lie less in groundbreaking new technologies and more in fostering a sustainable, mindful relationship with our environment. This approach doesn't offer a single solution or formula. For ecologically conscious artists, it might mean rethinking traditional manufacturing methods, accepting more limited palettes, or even working with simpler, less intense colours, as seen Chapter 2: The Science of Pigment.

Ultimately, it's about adapting to a way of living with colour that respects and asks questions about ecological boundaries, reimagining the role colour plays in our lives and creative practices.

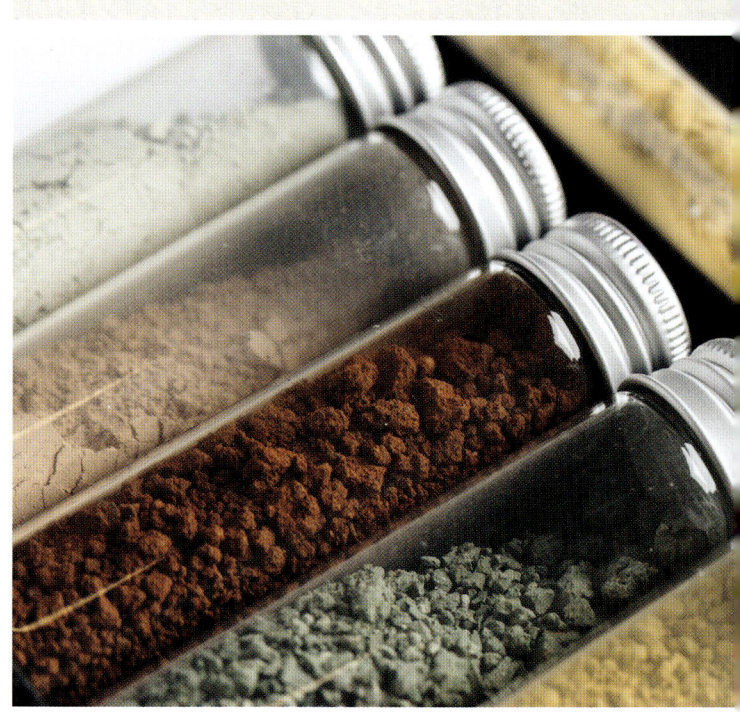

ABOUT THE AUTHOR

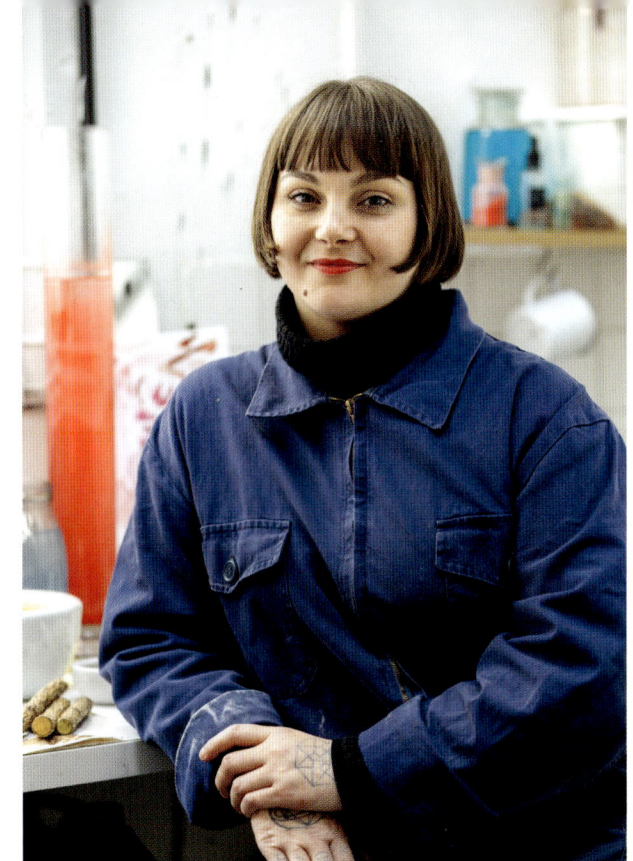

Lucy Mayes is an artist, pigment maker, and educator working in the UK between London and Hampshire. As London Pigment, she uses materials from urban waste streams to make recycled pigments for creative practitioners to use in their work. Her pigments and inks are available through art materials retailers and through her website. She studied Fine Art at Oxford University and Royal College of Art, and previously worked as pigment specialist at leading pigment retailer L. Cornelissen & Son. She has taught pigment making at the V&A, Kew Gardens, Tate, and Somerset House (amongst others). As a board member of non-profit Pigments Revealed International, she is passionate about sharing knowledge of the craft of pigment making.

To learn the craft of pigment making in person, contact Lucy via her website londonpigment.com or find her on Instagram @londonpigment.

ACKNOWLEDGEMENTS

I extend my deepest gratitude to all those who have supported me on my creative journey. Thank you to Nicholas Walt, as well as my esteemed colleagues and fellow makers, Keith Edwards and Magnus Sigurdsson Hardradi, at and affiliated with L. Cornelissen & Son.

I am also sincerely thankful to Melonie Ancheta and my past and present colleagues at Pigments Revealed International.

My appreciation goes to Jo Volley and Ruth Siddal for their support and for welcoming me into the cohort of the Materials Research Project at UCL Slade School of Fine Art.

I am grateful to the organizations that have supported my work, including Heritage Crafts, The British Plant Gall Society, The Royal Botanical Gardens at Kew, The Royal Cornwall Museum and West Dean College, and the V&A.

SPECIAL THANKS

To my family, my heartfelt thanks. To my parents, for raising a family of dreamers who have always pursued their passions. To my brothers, Adam and Simon, and my sister, Verity, for indulging my fascination with pigments and for their constant support. To Lloyd, for always standing by me and embracing my interests as his own.

I am also deeply appreciative of my artist friends for their unwavering encouragement, as well as to all those who have participated in my workshops, followed my work, shared it, or purchased my pigments.

To the global natural pigment and dyeing community for their generosity and support.

Lastly, my thanks to the art teacher who once told me I would win the Turner Prize – your belief has always stayed with me.

BIBLIOGRAPHY

Ball, P., *Bright Earth: Art and the Invention of Colour*, University of Chicago Press, 2003

Cennini, C., *The Craftsman's Handbook (Il Libro dell'Arte)*, Translated by D. V. Thompson, Dover Publications, 1933

Doerner, M., *The Materials of the Artist and Their Use in Painting: With Notes on the Techniques of the Old Masters*, Harvest Books, 1921

Eastaugh, N., et al, *Pigment Compendium: A Dictionary of Historical Pigments*, Elsevier Butterworth-Heinemann, 2004

Finlay, V., *Colour: A Natural History of the Palette*, Random House, 2002

FitzHugh, E. W., ed. *Artists' Pigments: A Handbook of Their History and Characteristics*, Volume 3, National Gallery of Art, 1997

Gettens, R. J. and Stout, G. L., *Painting Materials: A Short Encyclopaedia*, Dover Publications, 1966

Harley, R. D., *Artists' Pigments c. 1600-1835: A Study in English Documentary Sources*, Butterworth Scientific, 1970

Hodgskiss, T., 'Ochre use at Sibudu Cave and its link to complex cognition in the Middle Stone Age', *Azania: Archaeological Research in Africa*, Vol. 49, No. 4, 2014

Laurie, A. P., *The Painter's Methods and Materials*, Dover Publications, 1967

Loske, A., *Colour: A Visual History*, Thames & Hudson, 2019

Mayer, R., *The Artist's Handbook of Materials and Techniques* (5th ed.), Viking, 1991

Ormsby, BA., et al, 'British Watercolour Cakes from the Eighteenth to the Early Twentieth Century', *Studies in Conservation*, Vol. 50, No. 1, 2005

Thompson, D. V., *The Materials and Techniques of Medieval Painting*, Dover Publications, 2003

SUPPLIERS

UK

L. Cornelissen & Son
cornelissen.com

Jackson's Art
jacksonsart.com

Wallace Seymour Fine Art Products Ltd.
wallaceseymour.co.uk

EUROPE

Kremer Pigmente
kremer-pigmente.com

Sennelier
sennelier-colors.com

Zecchi Colori Belle Arte
zecchi.it

USA

Kremer Pigments Inc.
shop.kremerpigments.com/us/

Natural Pigment
naturalpigments.com

JAPAN

Pigment Tokyo
pigment.tokyo/

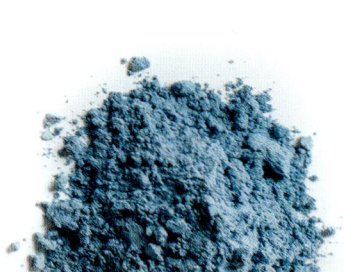
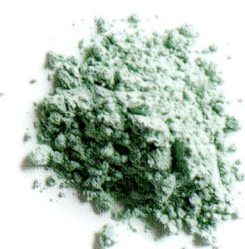

USEFUL ORGANIZATIONS

Pigments Revealed International
pigmentsrevealed.com

Heritage Crafts
heritagecrafts.org.uk

The British Plant Gall Society
britishplantgallsociety.org

Slade School of Fine Art Materials Research Project
ucl.ac.uk/slade/project/materials-research-project

REDiscover project
sites.google.com/fct.unl.pt/rediscover/

PIGMENT PEOPLE

Caroline Ross
foundandground.com/2018/10/27/found-and-ground/

Heidi Gustafson
earlyfutures.com/about/

Polly Bennett / POLBEN's PIgment
pollybennett.com/sustainable-practice

Melonie Ancheta
nativepaintrevealed.com/

Tilke Elkins / Wild Pigment Project
wildpigmentproject.org

Kauae Raro Research Collective
kauaeraro.com/about

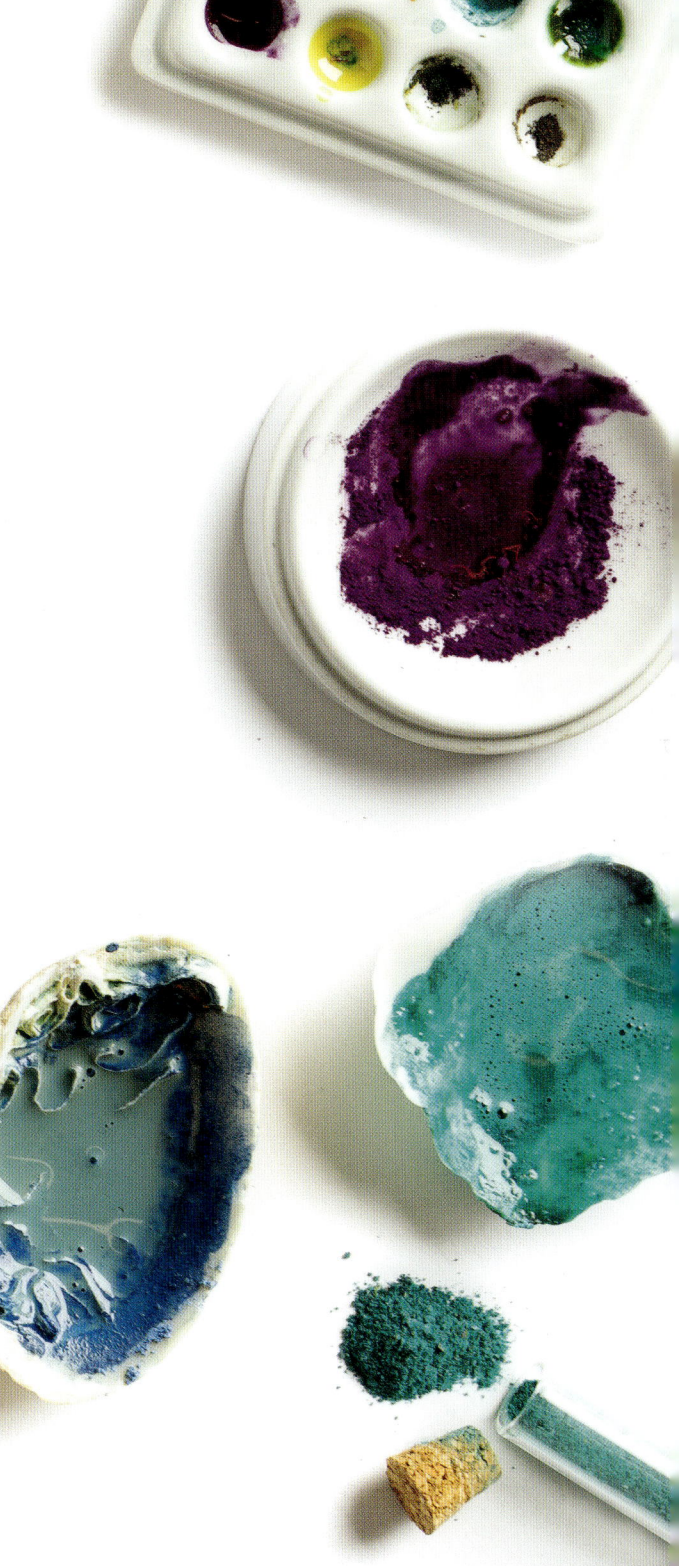
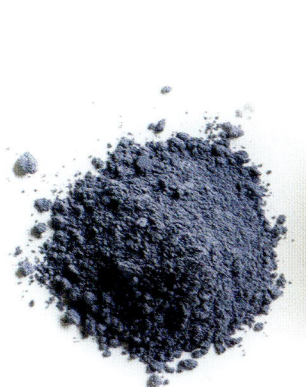
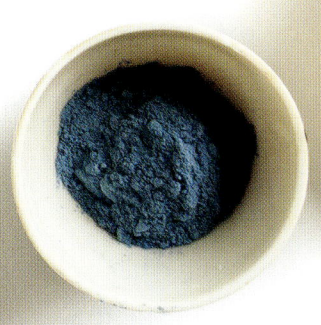

INDEX

ISBN-13: 9781446314814 hardback
ISBN-13: 9781446314906 EPUB

This book has been printed on paper from approved suppliers and made from pulp from sustainable sources.

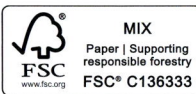

MIX
Paper | Supporting
responsible forestry
FSC® C136333

Printed in China through Asia Pacific Offset for:
David and Charles, Ltd
Suite A, Tourism House, Pynes Hill, Exeter, EX2 5WS

10 9 8 7 6 5 4 3 2 1

Publishing Director: Ame Verso
Publishing Manager: Jeni Chown
Senior Commissioning Editor: Nigel Browning
Editor: Victoria Allen
Development Editor: Miranda Harrison
Copy Editor: Cheryl Brown
Lead Designer: Sam Staddon
Designer: Anna Wade
Pre-press Designer: Susan Reansbury
Art Direction: Sarah Rowntree
Photography: Jason Jenkins
Production Manager: Beverley Richardson

David and Charles publishes high-quality books on a wide range of subjects. For more information visit www.davidandcharles.com.

Follow us on Instagram by searching for @dandcbooks.

Layout of the digital edition of this book may vary depending on reader hardware and display settings.

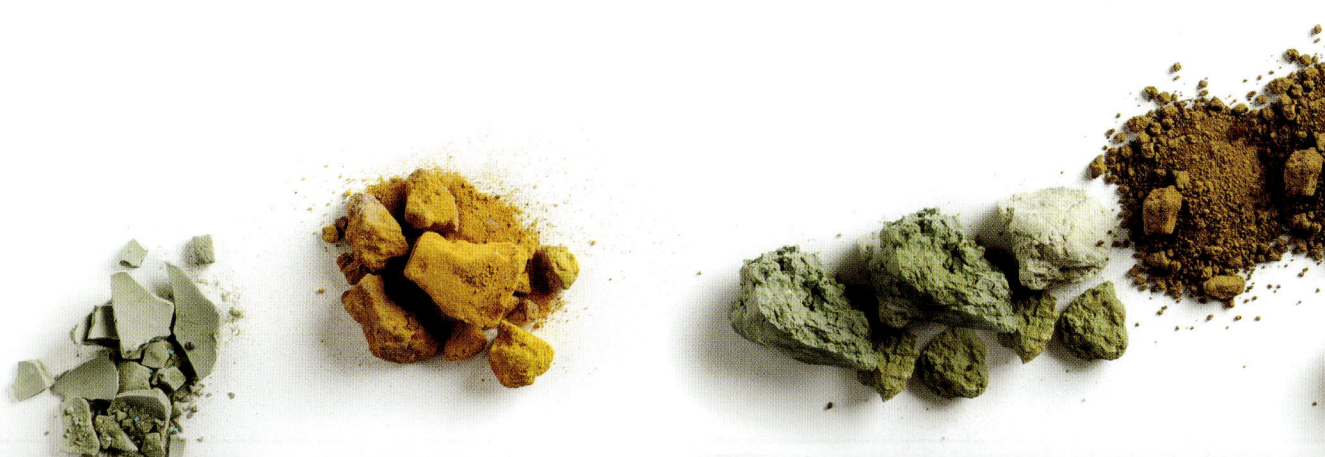

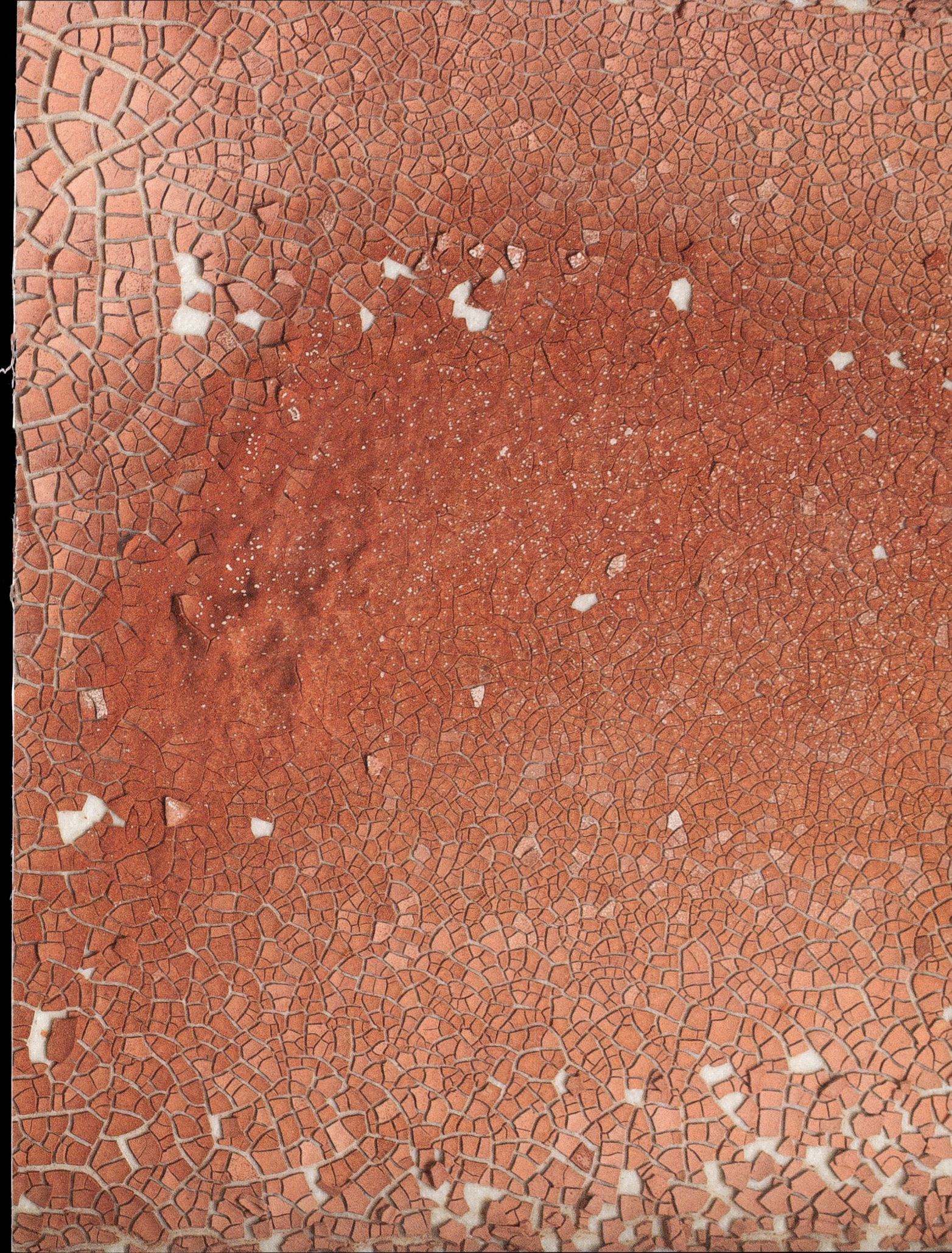

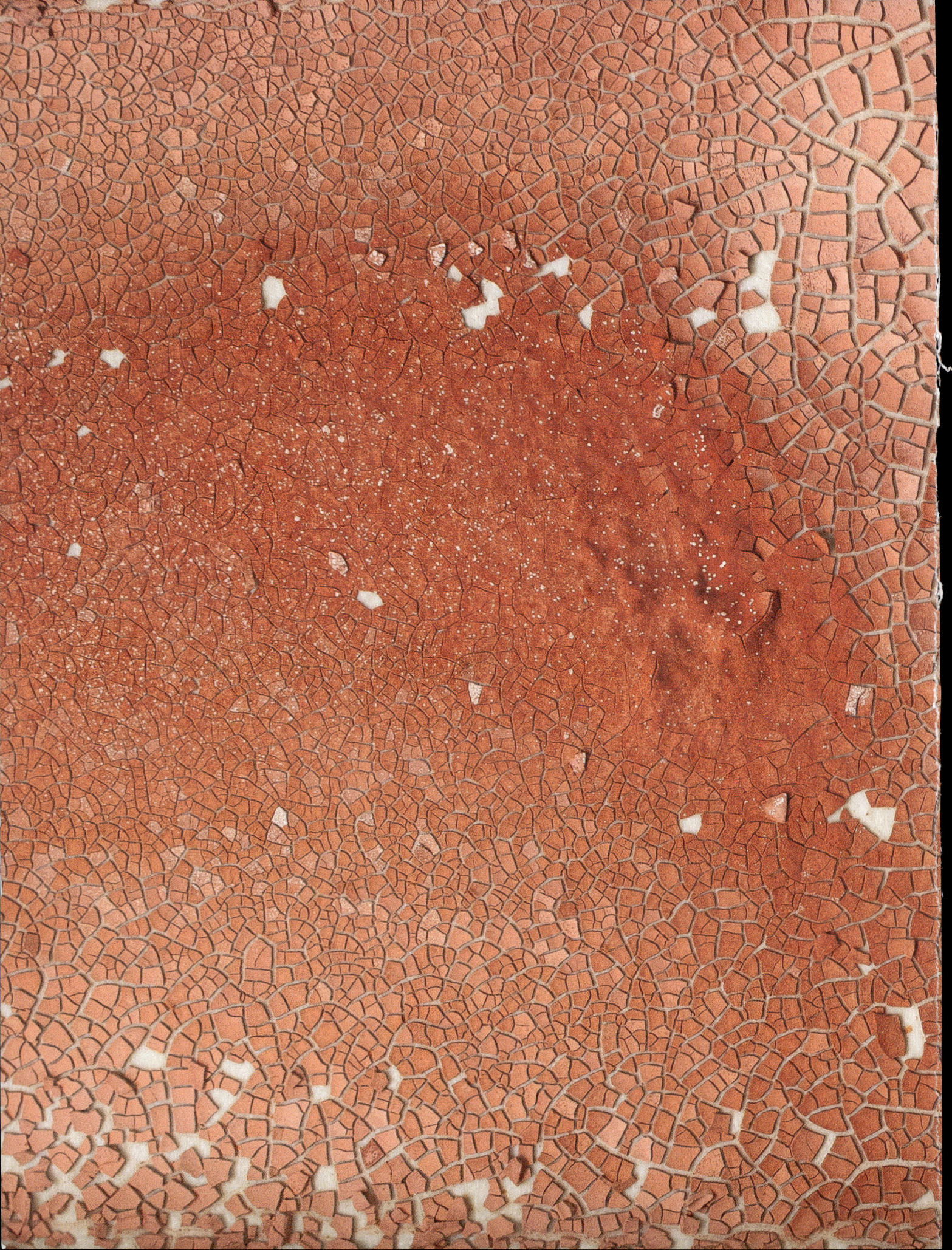